TELECOMMUNICATIONS LAW AND PRACTICE

AUSTRALIA AND NEW ZEALAND
The Law Book Company Ltd.
Sydney : Melbourne : Perth

CANADA AND U.S.A.
The Carswell Company Ltd.
Agincourt, Ontario

INDIA
N. M. Tripathi Private Ltd.
Bombay
and
Eastern Law House Private Ltd.
Calcutta and Delhi
M.P.P. House
Bangalore

ISRAEL
Steimatzky's Agency Ltd.
Jerusalem : Tel-Aviv : Haifa

MALAYSIA : SINGAPORE : BRUNEI
Malayan Law Journal (Pte.) Ltd.
Singapore and Kuala Lumpur

TELECOMMUNICATIONS LAW AND PRACTICE

by

COLIN D. LONG LL.B.

Partner in Bird & Bird

LONDON
SWEET & MAXWELL
1988

Published in 1988
by Sweet & Maxwell Ltd.
11, New Fetter Lane, London
Computerset by Promenade Graphics Ltd, Cheltenham
Printed in Great Britain by
Adlard and Son Limited,
Leatherhead, Surrey and Letchworth, Hertfordshire

British Library Cataloguing in Publication Data
Long, Colin D.
Telecommunications Law and Practice.
1. Telecommunications—Law and Legislation
—Great Britain
I. Title
344. 103'994 KD2880

ISBN 0–421–37780–1

To Sheila, Emily and Charles

I'll put a girdle round about the earth
In forty minutes

Puck, 'A Midsummer Night's Dream'

FOREWORD

The decade of the 1980s will be remembered as a time of remarkable change in British telecommunications. The change resulted from the implementation of a new political philosophy during a period in which technology was developing with unusual rapidity. The pivotal event has been the privatisation of British Telecom. This gave it the freedom from rigid controls and the incentives to operate more efficiently. It also necessitated the establishment of a regulatory regime, which is administered by the Office of Telecommunications (OFTEL). The focus of the regulations is to make things better for customers. As Director General of Telecommunications, I have the job of promoting the interests of customers partly by promoting competition. I regard competition as the regulator's best weapon: a monopolist can afford to neglect the customers, a business facing competition cannot. However, where competition is not yet effective, regulation is needed to prevent the abuse of monopoly power—and regulation is most beneficial if it mimics the effect of competition.

The new regime in telecommunications inevitably involves some complexity. Partly this is transitional. I aim to have as little complexity in regulation as possible and I expect that it can be reduced as time goes by. However, those who have described our policies as "deregulation" have chosen the wrong word. To promote competition, and to protect the interests of customers, in a situation where one supplier has a large degree of dominance, regulation is required and some of the regulations will involve legal—and other—complexity.

The development of a body of expertise about telecommunications law is therefore important to the success of the new policies. Colin Long is one of a group of lawyers who have specialised in this branch of law and, so far as I know, his book is the first in the field. No doubt, it, like British Telecom, will have to face competition. I welcome it and wish it success in contributing to—and participating in—effective competition.

9th June 1988

BRYAN CARSBERG
Director General of Telecommunications

PREFACE

Ten years ago there was, to all intents and purposes, no such thing as telecommunications law. Even in 1982, with the creation of Mercury Communications as a potential alternative carrier to BT, the scope for competition was being broadened but the mechanisms for its enhancement by regulation had not been developed nor even, perhaps, conceived.

The Government's ideals and its manifesto objectives were finally fully translated into practical and legal form and effect by the 1984 Act, but, even with the panoply of regulation introduced by that Act, the infrastructure for fair competition in both telecommunication service and apparatus supply left some gaps. In the absence of an American style of approach to anti-trust law no privatisation/liberalisation process of this kind can properly be implemented without in some way building on or modifying the pre-existing competition laws of this country. Clearly, in creating a new competition law "overlay" for telecommunications this point was recognised but even with the licensing regime now in place and the powers conferred upon the Director General and OFTEL under the 1984 Act, some weaknesses remain. The recent Government Green Paper on the proposed revision of United Kingdom restrictive trade practices policies shows that our laws regarding anti-competitive agreements will soon be brought into line with the effects-based approach of our continental colleagues and European law, but unfortunately the oppourtunity is not to be taken by Government to include similar new legislation controlling the abuse of a dominant position. The argument that the 1980 Competition Act effectively serves this purpose is not convincing.

The European Commission's Green Paper on the development of the Common Market for telecommunication services and equipment may have some impact, even in the relatively liberalised environment of the United Kingdom. Two directives are promised for 1988, as to apparatus and services; indeed, the opening up of national barriers to cross-frontier supply promises to be a big challenge to the Community, given the very national nature of each country's telecommunication undertakings, policies and regulations. Even the European Commission itself has not totally avoided the trap of traditional conservative thinking in this respect: the Green Paper acknowledges that network infrastructure, in terms of the physical apparatus necessary to "wire up" a country, should be the exclusive preserve of one undertaking but its suggestion that voice services should be similarly reserved rather cuts across the success of United Kingdom experience in opening up voice service to competition.

This book is designed to serve as an introduction to lawyers and businessmen and as an aid to more detailed consideration of the issues arising out of United Kingdom regulation of telecommunications. It is not intended as an out and out practitioners' textbook; I have therefore kept case law citations to a minimum, not that these are very much help in the area of licence regulations except by way of comparison with other regulated industries, such as electricity, where some of the social service concepts are similarly framed.

The licensing regulations, which form the bulk of the commentary in this book, should be less complicated if and when the restrictions on simple resale are removed next year, but the regulation of competition and therefore of service activity looks well set to continue as an issue into the 1990s, when further network competition may be allowed by the Government. Certainly, so long as BT remains a de facto monopoly, and United Kingdom regulation of competition generally fails to provide all the answers, some industry-specific controls over its activities will continue to be necessary.

In the immediate future there is clearly a need for

rationalisation and simplification of the regulations, particularly those in the Branch Systems General Licence and the Value Added and Data Services Licence. Whenever interpretation of the sometimes obtuse language of these regulations becomes necessary, one frequently has the feeling that if one could have been aware of the intention of the draughtsmen, the realisation of that intention in print would have been that much easier to comprehend.

Generally with United Kingdom regulation one is reminded of the story of the traveller who, on asking a bystander the way, received the response from the bystander that he would not have started here. If the opportunity ever arose to do away with the existing basis of regulation and move towards one where the running of telecommunications systems was in fact permitted from the outset and only specific acts or the provision of specific acts or the provision of specific services was prohibited, life would be much simpler.

At the time of finishing this book, there are many issues arising out of United Kingdom telecommunications regulation which remain unresolved or on which further pronouncements are promised. The meaning of many of the terms used in licences has also yet to be tested and it is thus inevitable that this book can, in some cases, only present the problems and possibilities without offering categoric answers. A very good example is the Director General's statement in the very recent PanAmSat* case in which he gives his view of the meaning of "reasonable demand" in typically simple and direct terms. In the coming years it can be expected that this phrase in particular will come under ever closer scrutiny in testing the extent of public telecommunication operators' obligations and responsibilities to their customers. After all telecommunications law is in the end about the rules introduced to promote the interests of customers and not simply to create new business activity.

ACKNOWLEDGEMENTS

I am indebted to a number of people for their assistance to me in commenting upon the book. First of all to Tessa Dunstan, the former Legal Adviser to the Director General of Telecommunications, to John Roberts, her successor, and Eily Harty, Assistant Legal Adviser at OFTEL, to Paul Franklin at EDS, Michael Corby of Commed, Maev Sullivan (Legal Adviser on regulatory affairs at Mercury Communications Limited) and my colleague and partner, David Kerr.

My particular thanks to Stephen Kingsley for his contribution of Chapter 10 without which this book would have been incomplete.

I must also record how grateful I am to a number of clients in the telecommunications industry, in particular Mercury, as but for the privilege of working for them over the years and gathering an invaluable awareness and experience of a fascinating industry, this book would never have been written.

Finally, my special thanks to my wife Sheila, without whose encouragement, patience and support the writing of this work would have been a much more difficult, if not painful, process.

The book was substantially complete at the end of 1987 and therefore the laws and regulations are, to the best of my belief, up to date at that point. Subsequently, I had the opportunity to incorporate references to certain new developments, such as the Director General's PanamSat decision, but such is the pace of evolution in this area that it is impossible to guarantee that developments in the early part of 1988 have all been covered. Indeed, there has to date been so little opportunity for practical or judicial testing of the rules and procedures discussed in this book that it must always be preferable for reliance to be placed on specific advice tailored to a given situation.

CONTENTS

GLOSSARY

ACT	Advisory Committee on Telecommunications.
1973 Act	The Fair Trading Act 1973.
1980.Act	Competition Act 1980.
1981 Act	The British Telecommunications Act 1981.
the Act	Telecommunications Act 1984.
Administration	A body or organisation responsible for running a national telecommunications system and, in many cases, for administering rules as to the operation of customer systems and the attachment of terminal apparatus. See also PTT.
Analogue	A type of telecommunication system. See Chapter 1, section 1.4.
ATIEP	Association of Telephone Information and Entertainment Providers.
BABT	British Approvals Board for Telecommunications.
BSGL	Branch Systems General Licence.
BSI	British Standards Institution.
BT	British Telecommunications plc.
Call Routing Apparatus	Apparatus (such as a PABX) switching 2-way live speech calls between two or more extensions and two or more incoming circuits. See Condition 1 of General Conditions applicable to Class Licences.
CBA	Cable & Broadcasting Act 1984.
CCA	Consumer Credit Act 1974.
CCITT	Comité Consultatif International Telegraphique et Telephonique

	(International Telegraphic and Telephonic Consultative Committee).
CEPT	European Conference of Post and Telecommunications Administrations.
Copyright Act	Copyright Act 1956.
CPA	Consumer Protection Act 1987.
DBS	Direct Broadcasting by Satellite.
Designated Maintainer	A person providing maintenance services for Call Routing Apparatus, as defined in Condition 1 of the General Conditions applicable to Class Licences.
Digital	A type of telecommunication system communicating in binary digits in the same way as a computer. See also Chapter 1, section 1.4.
Director General	The Director General of Telecommunications.
DGFT	Director General of Fair Trading.
DTI	Department of Trade and Industry.
FSS	Fixed Service Satellite.
IPC	International Private Circuit.
ISDN	Integrated Services Digital Network.
ITU	International Telecommunications Union.
Major Service Providers	Licensees under the VADS licence who exceed the turnover limits set in Condition 5 (m) of that licence.
Mercury	Mercury Communications Limited.
MMC	Monopolies and Mergers Commission.
Modem	Apparatus for combining different signals for them to be conveyed together over the same communications channel.
Multiplexing	Equipment which modulates the signal to be transmitted and

	demodulates the signal received, used for converting digital signals into an audible tone suitable for analogue circuits; see also Chapter 1, section 1.4.
NETS	Normes Européennes de Telecommunications.
Node	A junction point in a network, public or private, providing the means of access to or egress from the network.
NTP	Apparatus at the boundary of a telecommunication system, forming the interface with another system. See Annex A paragraph 2 (i) to the BT licence.
NTTA	Apparatus of a PTO installed on customer premises, incorporating the NTP of the PTO and customer systems and providing the means of physical connection and disconnection, as well as testing (see further Annex A paragraph 2 (j) BT licence) of each system.
OFTEL	Office of Telecommunications.
OSI Standard	Open Systems Interconnection. See for example BT Licence Condition 40A.
PABX	Private Automatic Branch Exchange.
Packet switching	A method of conveying data, divided into packets of information, further explained in Chapter 7, section 2.4.
PSN	Public switched network.
PSTN	Public switched telephone network.
PTO	Public telecommunications operator, licensed under Section 7 of the Act and so designated under Section 9 of the Act.
PTO Associates(s)	A member of a PTO's corporate

	group—see for example PTO Associates VADS licence, Chapter 6, section F.
PTT	Post, Telegraph and Telephone, the generic description of a public utility running a country's telephone network.
RPOA	Recognized Private Operating Agency.
Secretary of State	The Secretary of State for Trade and Industry.
SGSA	Supply of Goods and Services Act 1982.
Simple Resale	The use of a private circuit to convey messages to and from the PSN.
SITS	Special Investigation Test Schedules.
SMATV	Satellite Master Antenna Television.
Supplemental Services Business	The business of a PTO equivalent to a value added services business capable of being run by a non-PTO under the VADS licence.
Systems Business	A PTO's business of providing basic telecommunication services and ancillary activities, for example, defined in Condition 18 of BT's licence.
TACS	Telecommunications Advisory Committees.
Telecommunications Code	The Telecommunications Code as contained in the Act. See also Chapter 10.
TIS	Technical Information Sheet. See Chapter 8.
UCTA	Unfair Contract Terms Act 1977.
VADS	Value Added and Data Services.
VADS Licence	The Class Licence for VADS.
VANS	Value Added Network Services.
WTA	Wireless Telegraphy Act 1949.

TABLE OF CASES

A-G *v.* Edison Telephone Co. of London (1880) 6 Q.B.D. 244 2
Allan (Merchandising) Limited *v.* Cloke [1963] 2 Q.B. 340 35
Archbolds (Freightage) Limited *v.* S. Spanglett Limited [1961] 1 Q.B. 374 35
Associates Provincial Picture Houses Ltd. *v.* Wednesbury Corporation [1947] 2 All E.R. 680 20

Brinkibon *v.* Stanag Stahl [1983] 2 A.C. 34 ..... 185, 187

Carlill *v.* Carbolic Smoke Balls [1893] 1 Q.B. 256 at 262 185
Coditel *v.* Cine Vog Films [1981] 2 C.M.L.R. 362 173
Council of Civil Service Unions *v.* Minister for the Civil Service [1984] 3 All E.R. 935 20

Dungate *v.* Lee 1969 1 Ch. 545 35

Entores Limited *v.* Miles Far East Corporation [1955] 2 All E.R. 493 ..... 185, 188

Hadley *v.* Baxendale (1854) 9 Exch. 341 180
Henkel *v.* Pape (1870) C.R. 6 Exch. 7 188

Italy *v.* European Commission [1985] C.M.L.R. 368 144

Levy *v.* Yates 1838 S.A. & E. 129 35

Preston *v.* I.R.C. [1985] 2 All E.R. 327 20

R. *v.* Gold & Schifreen 177

South of Scotland Electricity Board *v.* The British Oxygen Company [1956] 1 W.L.R. 1069; [1959] 2 All E.R. 225 60

Victoria Laundry (Windsor) Ltd. *v.* Newman Industries Ltd. [1949] 2 K.B. 528 181

Wheeler *v.* Leicester City Council [1985] 2 All E.R. 1106 20

TABLE OF STATUTES

1863 Telegraph Act (26 & 27 Vict. c. 112) ... 152
1868 Telegraph Act (31 & 32 Vict. c. 110) ... 2, 3, 152
1878 Telegraph Act (41 & 42 Vict. c. 76) 152
1892 Telegraph Act (55 & 56 Vict. c. 59) 152
1899 Telegraph Act (62 & 63 Vict. c. 38) 152
1908 Telegraph (Construction) Act (8 Edw. 7, c. 33) 152
1909 Telegraph (Arbitration) Act (9 Edw. 7, c. 20) 152
1911 Telegraph (Construction) Act (1 & 2 Geo. 5, c. 39) 152
1916 Telegraph (Construction) Act (6 & 7 Geo. 5, c. 40) 152
1947 Electricity Act (10 & 11 Geo. 6, c. 54)—
s. 37 (8) 60
1949 Wireless Telegraphy Act (12, 13 & 14 Geo. 6, c. 54) 116, 149, 175
s. 1 (4) 30
s. 5 179
s. 19 28
(1) 29
1950 Public Utilities Street Works Act (14 Geo. 6, c. 39) 159
1952 Defamation Act (15 & 16 Geo. 6 & 1 Eliz. 2, c. 66)—
s. 9 174
1956 Copyright Act (4 & 5 Eliz. 2, c. 74) 166, 172
s. 14 .... 169, 172
s. 14A 169, 172
1967 General Rate Act (c. 9)—
s. 21 158
1971 Town and Country Planning Act (c. 78)—
s. 52 162
1973 Water Act (c. 37) 61
Fair Trading Act (c. 41) 18, 23, 140, 141, 145
Pt. III 22, 23
ss. 47–56 141
s. 48 142
s. 55 (1) 142
s. 64 143
s. 69 (2) 62
s. 70 143
s. 75 62
s. 78 142
Sched. 8, Pt. I 141, 142
Pt. II 141, 142
1974 Local Government Act (c. 7) 158
s. 19 158
Consumer Credit Act (c. 39) 184
1976 Water Charges Act (c. 9) 61

1977 Unfair Contract Terms Act (c. 50) ... 181
s. 2(1) ... 182
s. 12 ... 181
1979 Sale of Goods Act (c. 54) ... 183
1980 Competition Act (c. 21) ... 18, 23, 140, 141, 143, 145
s. 2 ... 144
ss. 2–10 ... 140, 143
s. 7(6) ... 143
1981 British Telecommunications Act (c. 38) ... 9, 10, 12, 16, 25, 41, 42, 96
s. 3 ... 12
Forgery and Counterfeiting Act (c. 45) ... 177
Acquisition of Land Act (c. 67) ... 160
1982 Supply of Goods and Services Act (c. 29) ... 180, 183, 184
s. 13 ... 180
1984 Telecommunications Act (c. 12) ... 10, 12, 16, 20, 92, 97, 116, 122, 125, 138, 149, 151, 152, 165, 176
Pt. II ... 16, 18, 31
Pt. III ... 16, 18, 46
s. 1 ... 16, 37
s. 3 ... 16, 49, 61, 182
(1) ... 17
(2) ... 17, 61
(3) ... 22
s. 4(1) ... 32, 33, 35
(2) ... 33, 34
(3) ... 35, 50, 52, 100
(7) ... 47, 99
ss. 5–11 ... 18
s. 5 ... 31, 32, 34, 126
(1) ... 37

1984 Telecommunications Act—*cont.*
s. 5—*cont.*
(2) ... 32, 35, 37
(4) ... 34
s. 6 ... 30, 35, 44
(1) ... 116
(3) ... 35
(4) ... 35
s. 7 ... 29, 30, 31, 38, 40, 41, 46, 134, 141
(6) ... 36, 106
s. 8 ... 13, 42, 46
s. 9 ... 13, 42
s. 10(2) ... 43
ss. 12–15 ... 18, 141
s. 12 ... 38
(4) ... 38
(6) ... 39
(*a*) ... 39
s. 13 ... 37, 39
(5) ... 39
(9) ... 39
ss. 16–18 ... 26
ss. 16–19 ... 18, 138
s. 16 ... 38, 46
(7) ... 138
s. 18 ... 139
(1) ... 139
(6) ... 38, 140
(7) ... 139
s. 19 ... 18
s. 20 ... 18, 89, 135
s. 21 ... 18
s. 22 ... 18, 56, 72, 89, 125, 126
(1) ... 13
s. 23 ... 18
s. 28 ... 130
s. 29 ... 130
s. 31 ... 158
s. 34 ... 160
s. 42 ... 177, 178
s. 43 ... 123, 178
s. 44 ... 176
s. 45 ... 176, 177
s. 47 ... 18
s. 48 ... 18

1984 Telecommunications Act—*cont.*
s. 49 .... 18, 140
s. 50 .... 18, 22, 140, 141
(1) 22
(2) 23, 141
(3) 23
s. 58 47
s. 59 (4) 141
s. 74 30
s. 95 .... 39, 141
s. 96 162
s. 106 (2) 32
Sched. 2 .... 43, 151
(Telecommunications Code) 13, 43, 45, 46, 151, 153, 165
para. 2 .... 160, 162
(7) 159
para. 3 .... 160, 162
para. 4 162
para. 5 .... 160, 163
para. 7 161
para. 8 163
(5) 164
para. 9 158
para. 10 161
para. 12 161
para. 16 162
para. 17 163
para. 18 162
para. 19 161
para. 20 164
para. 21 .... 161, 164
para. 23 159
para. 26 161
1984 Telecommunications Code—*cont.*
Sched. 5, para. 1 41
(2) 41
para. 12 180
Cable and Broadcasting Act (c. 46) 120, 122, 149, 166, 171, 172
s. 2 (1) 118
(3) 119
s. 4 .... 47, 118
s. 5 119
s. 6 119
s. 7 119
s. 28 174
1985 Companies Act (c. 6) 86, 97
s. 736 .... 86, 108
Copyright (Computer Software) Amendment Act (c. 41)—
s. 2 169
Interception of Communications Act (c. 56) 174, 176
s. 1 175
s. 2 .... 175, 176
1986 Insolvency Act (c. 45)—
s. 233 183
1987 Consumer Protection Act (c. 43) 184
Pt. II 184
s. 2 184

TABLE OF STATUTORY INSTRUMENTS

1960 Plant and Machinery (Rating) Order (S.I. 1960 No. 122) 158
1965 Rules of the Supreme Court (S.I. 1965 No. 828)—
Order 11 185
1977 Town and Country Planning General Development Order (S.I. 1977 No. 289) 157
Art. 4 156
Class I 156
Class XVIII 152, 153
Class XXIV 153, 154, 155, 156
Class XXV 153, 155, 156
1980 Anti-Competitive Practices (Exclusions) Order (S.I. 1980 No. 979) 144
Sched. 1 144
1984 Cable Programme Services (Exceptions) Order (S.I. 1984 No. 980) 122
1985 Town and Country Planning (National Parks, Areas of Outstanding Natural Beauty, and Conservation Areas, etc.) Special Development Order (S.I. 1985 No. 1012) ... 156
Telecommunication Apparatus (Marketing and Labelling) Order (S.I. 1985 No. 17; amended by S.I. 1981 No. 1031) ... 130
Telecommunication Apparatus (Advertisements) Order (S.I. 1985 No. 719; amended by S.I. 1985 No. 1030) 130
1986 Telecommunications Act (Extension of Relevant Period) Order (S.I. 1986 No. 1275) 41

Chapter 1

INTRODUCTION

A. The World of Telecommunication

1. General History

Telecommunication, literally communication at a distance, may have had humble beginings (semaphore and smoke signals are two of the most basic examples) but modern technology and particularly the digital language of computers has transformed the art of telecommunication into a science of breathtaking speed and complexity.

As in its origins, the most familiar manifestation of telecommunication today remains the telephone instrument itself. Alexander Graham Bell is credited with invention of the telephone but, like many epoch-making inventors, the fact that he was first was a combination of opportunism and timing as well as of skill and persistence. The beginnings came in the early part of the nineteenth century: in the 1820s an English Scientist Charles Wheatstone, demonstrated that musical sounds could be transmitted through metallic and glass rods. Wheatstone was later involved, in 1837, in the invention of the first electrical telegraphic transmission device but failed, along with a number of his contemporaries, to appreciate the potential of some of the instruments he had designed. Bell it was who, in 1875, appeared to have been the first to understand the commercial implications of the electrical transmission of voice.

In the period 1872 to 1875 Bell, in parallel with Elisha Gray, both working in the United States, was nearing the culmination of his efforts to produce a working telephone; both men worked feverishly towards the same goal but in different ways encountered their various setbacks. Bell seemed to have the greater confidence in his invention and certainly had the greater commercial acumen. On February 14, 1876 Gray filed with the United

States Patent Office a caveat, or notice of invention. This was not enough. On the same day Bell filed his patent application. At this point, surprisingly, neither man had successfully transmitted speech, but this was soon to come. On March 10, 1876, three days after Bell's patent was issued, he successfully transmitted words using a variable resistance transmitter. Soon afterwards Thomas Edison designed a carbon resistance transmitter and it was this type of transmitter that was destined to remain the basic standard into the twentieth century.

Thereafter the pace of development of the sophistication and usage of the telephone accelerated rapidly. By the mid–1880s the switchboard had evolved to a point where hundreds and even thousands of subscribers could be handled. Private companies in London and in a number of provincial towns introduced telephone exchanges and the Post Office's telegraph "control switchings" in several towns were converted to telephone exchanges.

2. United Kingdom History

In England, as entrepreneurial activities in the establishment of telephone systems began to develop and necessarily impinge on the Post Office's telegraph monopoly, the question arose as to whether such activities amounted to an infringement of the Postmaster-General's "exclusive privilege" of transmitting telegrams, as conferred by the Telegraph Act of 1869. The issue came before the High Court, Exchequer Division, in 1880 and the recital of facts in the case gives a fascinating insight to the workings and social impact of the original telephone.[1] The agreement made by the Edison Telephone Company with its customers (to whom it leased the necessary equipment) provided that the company would:

> "Upon the request made through the said telephone at any time during the continuance of this agreement, between the hours of 9 a.m. and 6 p.m., Sundays excepted, put the lessee in telephonic

> communication with the telephone of any other subscriber to the said exchange whose wire is free."

(Such a guarantee of connection provides an interesting contrast with the liability accepted by PTOs under current standard conditions of contract).

The Court found that the scope of the 1869 Act was such that any apparatus for transmitting messages by electric signals must be a telegraph whether a wire was used or not. This would extend to "electric signals made, if such a thing were possible, from place to place, through the earth or the air." A somewhat prophetic statement.

The Edison Telephone Company argued unsuccessfully that the telephone and telegraph were significantly different technically and that the Postmaster-General's exclusive privilege should not extend to something not even invented in 1869. The Court held that a conversation through a telephone represented a message or communication that was in fact transmitted by a telegraph.

Thus was confirmed the exclusive privilege of the Postmaster-General over the telephone, a privilege which was to continue for over a century until 1981. By that time it had been refined as "the exclusive privilege of running systems conveying speech, music and visual images by electric, magnetic or similar means." The Post Office's (by then British Telecom's) monopoly of telecommunications had been extended to the setting of technical specifications and the provision and maintenance not only of telephone instruments and associated wiring in private premises, but also of any private branch exchanges connected to the public network.

Gradually after the Edison case the private companies, which in the late 1800s developed the infrastructure of wires, systems and exchanges much as we have today, merged or went out of business, selling their undertakings to the Post Office, so that by 1913 the Post Office had become the monopoly supplier of telephone services in all areas of the United Kingdom except the City of Kingston-upon-Hull, which retained its licence to operate its own municipal system and still does today, very successfully.

The first automatic telephone exchange was opened in Epsom in 1912, by which time there were some 700,000 telephone customers in Great Britain. By 1923 the one million mark was passed and there are, currently, some 20 million or more exchange connections. In the United Kingdom in 1984, the year of BT's privatisation, 23 thousand million inland telephone calls and 173 million international telephone calls were made. The percentage of United Kingdom households with a telephone is now over 80.

Today's telephone calls no longer simply involve one human being talking to another, least of all in the same language. Voice traffic still accounts for the lion's share of services but traffic in data is increasing much faster. Mobile communication services based on wireless telegraphy are bounding ahead; it could be that within a few years the mobile telephone instrument will do to the wired variety what the transistor radio has done to the valve-powered "wireless" set. On the back of the new technologies a range of computer-controlled communication services has evolved, coining the generic title "information technology."

3. Importance of Telecommunications to the Economy

Effective communication is as important to a nation's economy as it is to a modern mobile army. In November 1981, the General Assembly of the United Nations adopted a Resolution which stressed "the fundamental importance of communications infrastructures as an essential element in the economic and social development of all countries." The EEC Green Paper "Towards a Dynamic European Economy" (Development of the Common Market for Telecommunication Services and Equipment)[2] proclaimed that:

> "The emergence of an advanced and efficient European telecommunications system will cause deep

changes in the economy: more efficient organization of production, narrowing differences in geographic location, growing efficiency/innovation of services etc."

The first and most obvious way, therefore, in which this importance manifests itself is in the greater speed and efficiency with which all peoples and their institutions can go about their day-to-day lives and activities. Then there is the vital contribution that can be made to a domestic economy by an indigenous telecommunication equipment manufacturing base. Sales of such equipment by the United Kingdom industry exceeded 1.4 billion in value in 1984; approximately 12 per cent. of such sales were for export, yet the United Kingdom's sadly diminishing share of world exports is less than five per cent. World spending on telecommunications equipment in 1987 is likely to have exceeded $100 billion [Telecommunications Industry Research Centre—1987 World Outlook], with the United Kingdom market currently ranked seventh in the world.

4. Digital and Analogue Techniques

Much mention will be made in this book of digital and analogue systems. Currently most existing telephone networks are based on analogue systems. These turn the air pressure waves created by speech into analogous, constantly varying, electrical waves and turn them back to speech again at the receiver. By contrast digital systems convert speech, text, data and video into a stream of bits represented in the form of pulses of electricity or light. Further, by operating at high speed, it is possible to interleave the pulses relating to separate messages, such as different telephone calls, along a single line, a technique known as time division multiplexing.

All telephone exchanges were manually operated until 1912 when the electro-mechanical Strowger switching system was first introduced into the United Kingdom and became the standard exchange equipment.

5. Stored Programme Control Switching Systems

The most recent development in exchange technology has been the introduction of digital switching equipment incorporating stored programme control. Here the instructions to operate the exchange are contained in software programs which run on processors similar to data processing computers. Electro-mechanical switching equipment has thousands of moving parts which are costly to maintain and cause faults such as crossed lines, crackles and wrong numbers, whereas digital switching equipment contains almost no moving parts. Furthermore the power and performance of such equipment is very much governed by the software which it incorporates, so that improvements can be made to the equipment by the simple replacement and upgrading of software modules. At the same time multiplexers can vastly increase capacity carried over lines to which they are connected.

6. Optical Fibre Systems

New transmission links are for the time being to be made by the use of optical fibre cable, a medium allowing communication by pulses of light along tiny fibres of glass and susceptible of ever-increasing capacity as more sophisticated equipment is harnessed to it. These fibres are so pure that very little energy is lost as the signals pass through them and fewer repeaters are necessary. A single fibre less than 0.1 millimetres in diameter can carry 2,000 or more simultaneous conversations, whereas the previous generation of systems based on coaxial cables, with far less capacity, had cables of about 10 millimetres in diameter.

7. Computers and Data

The use of computers is creating new services and markets in telecommunications which were unprecedented and unthought of even 10 years ago. Any description of the structure of these markets is likely to prove as readily outdated as the services themselves. However, at present the major markets which seem set to fuel the continuing boom in telecommunication activities are inter-office voice and data communication over both public and private networks, value added and mobile services. Modern offices consist of a whole range of different machines for facsimile transmission, telex, view-data and other applications through switches (PABXs) on the premises or remote from the premises (*e.g.* Centrex). These machines can transmit their messages for switching and where appropriate, reprocessing, by computers embodied in public, and in many countries, private networks, before finally arriving at their destination. To aid in this process, and in advance of the complete digitisation of public networks, modems can be attached to customer terminal equipment to convert digital computer data into analogue signals, for transmission over analogue systems.

8. Private Networks

Private networks are composed of point-to-point lines leased from the public operators and connected together by microprocessor-controlled equipment. These effectively bypass the PTO's public switched network and (subject to domestic rules and international recommendations as to resale of capacity) allow the lessee to provide services to customers which, in many cases, will compete with the PTO's own services. The major problem faced by private network operators to date is not in so much dealing with the PTOs, but building a network that effectively links together computer based equipment produced at different times by different manufacturers using different standards. The need to overcome this hurdle if the momentum of telecommunications growth is to be sustained has been recognised in the move, started by the manufacturers and now enthusiastically encouraged by the regulators, towards open network standards or, in the industry vernacular, "open systems interconnection."

However, the progress of such harmonisation may be slower than the regulators first envisaged.

The increasing proliferation of private networks and the technical problems of intercommunication with the public networks has also brought about the requirement for common design criteria in order that acceptable call quality should be maintained. In December 1986 OFTEL published its "Provisional Code of Practice for the design of private telecommunication branch networks"[2] specifying the relevant design criteria. The code deals with telephony and covers such matters as call path transmission quality, call progress indication and numbering and is to be supplemented by other pronouncements from OFTEL on data networks.

9. ISDN

Common standards will be particularly important for the introduction of integrated services digital networks or "ISDN." This is the realisation of the present dream of all communications equipment channelling their messages in binary form over a common transmission facility, to be "read" processed and switched not only by the digital equipment at each end, but also by intelligent equipment within the public networks. With ISDN the standard telephone line will be transformed into a communications "pipe" with a digital capacity of 144 kbps, divided into two 64 kbps "bearer" circuits and a 16 kbps circuit for control information. Over this facility, digitised voice and data messages will be able to be pumped into and out of all manner of different, but integrated, systems performing different services. Nationwide ISDN service is being touted as likely to be available from 1990.

10. Standards

Progress in the integration and interconnection of value added and data services will equally be dependent on the introduction of common standards. These services can include electronic mail, remote processing of data, information retrieval, voice messaging, electronic banking and inventory management. Accessed over the fixed links provided by public networks, these services offer something over and above what is known in the United Kingdom as "basic conveyance" of messages.

These services are now regulated by one or other of two "class" licences.[4]

B. LIBERALISATION AND PRIVATISATION IN THE UNITED KINGDOM

1. British Telecommunications Act 1981

As has been mentioned, in the United Kingdom the exclusive privilege of the Post Office over telegraph and telecommunication continued until 1981. The British Telecommunications Act of that year made the postal and telecommunications services of the Post Office the responsibility of two separate undertakings. Postal services were retained by the Post Office but the Act transferred to British Telecommunications the exclusive privilege of running telecommunication systems. Having given with one hand, however, the 1981 Act gave Government the opportunity to take away with the other: this exclusive privilege was said not to be infringed by anything done under a licence granted by the Secretary of State. The political decision had by that time been taken to use this power to authorise one additional operator to compete with BT and a private joint venture company, Mercury Communications Limited (now wholly owned by Cable & Wireless plc) was established to operate a new public voice and data system based entirely on digital technology.

The 1981 Act also empowered the Secretary of State to permit persons other than BT to supply and maintain PABX equipment, although in practice this did not start to become effective until the spring of 1983.

At the same time the Secretary of State's power to licence the running of new telecommunication systems was utilised in order to allow the provision of value added network services or "VANS" over the public network.

Subsequently licences were also granted to two operators of the first ever cellular radio mobile telephone systems, Racal Vodafone and Telecom Securicor ("Cellnet").

2. Telecommunications Act 1984

Even as the 1981 Act was reaching the statute book, plans were being made for a much more radical approach, accomplished by the Telecommunications Act 1984, and involving the privatisation of BT by the sale of just over 50 per cent. of its equity, the abolition of its exclusive privilege and its replacement by a public operator's licence, as well as the termination of its right to grant licences, with in addition increased liberalisation by widening the categories of licensed system operators, reducing the involvement of BT in the approval of apparatus and establishing a competitive environment administered by a "watchdog" body (OFTEL) which would attempt to regulate and promote fair and effective competition and benefit consumers.

In contrast to the United States, where liberalisation entailed the break-up of AT&T and its Bell local operating companies, the BT privatisation was to leave BT and its *de facto* monopoly intact. One of the main reasons advanced for this is that although BT should experience competition at home, it should be left in as strong a position as possible to compete in international markets. Any dismembering of BT, so the argument would run, might erode its basic strength and prejudice its international competitiveness.

With the benefit now of hindsight it can be argued that the AT&T approach might have worked just as well in the United Kingdom. Hull City Council runs a very successful local telephone company and there is no reason why other major conurbations could not have had their own. Cable television operators could in theory provide local competition in telephony as well as video and interactive services, but so far progress in this direction has been tentative.

3. Government Policy for Competition in Telecommunication Services

If businessmen were to devise a strategy for the liberalising market in telecommunications and to be able to make the long-term investment decisions which were at the heart of this process, they needed to have some certainty, albeit subject to the whims of the electorate, as to the Government's policy in this area. In particular the flotation of BT and the expansion of Mercury's activities would not have been possible without the competitive context being clearly outlined. In order therefore to set the stage for the Act and to meet these calls for clarity, on November 17, 1983 the then Minister for Information Technology, Kenneth Baker M.P., made a statement on future competition policy to the Standing Committee on the Telecommunications Bill.[5] The full text of this is set out in Appendix A but its fundamentals were:

(a) The Government indicated that it did not intend to licence operators other than BT and Mercury to provide the basic telecommunication service of conveying messages over fixed links, whether cable, radio or satellite, both domestically and internationally, during the period to November 1990.

(b) Public systems were to be interconnectable: the Government's stated intention was that any subscriber to one public telecommunication system should be able to call any subscriber to any other public system. The licences of the operators of such systems would provide the rights and obligations for such interconnection and in particular BT's licence would oblige BT to connect its system to any other system where the operator of such system was licensed to make such connection. Where the parties could not agree on the terms of such connection these were to be determined by the Director General. (These provisions now appear in Condition 13 of BT's licence; for a fuller discussion of the scope of

Condition 13 and the first determination issued by the Director General under that Condition see below.)

(c) Simple Resale[6] of capacity over circuits leased from BT and Mercury would not (with certain exceptions, particularly now in relation to data services) be permitted in the period before July 1989, which corresponded to the original period (which may be extended) in BT's licence for application of the RPI–3 formula.[7]

4. The Director General and OFTEL

To some extent the 1981 Act was a mere dress rehearsal of liberalisation. With the Act and the privatisation of BT, including the abolition of its exclusive privilege, the serious work had begun. The Act provided for the new appointment of a Director General of Telecommunications, who was in fact first appointed by the Secretary of State for a three year period from July 1, 1984. The first Director General is Professor Bryan Carsberg, formerly a Professor of Accounting at London University. On July 1, 1987 his appointment was renewed for a further five years until June 30, 1992.

Under the Act the Director General is permitted to establish a Directorate known as the Office of Telecommunications which advises him on the exercise of his duties. Those duties assigned to him by the Act are set out in section 3. Essentially he is to exercise his powers so as to secure the provision of telecommunication services in order to satisfy all reasonable demands and so as to secure that the persons providing such services are able to finance them. More specifically he is bound to promote the interests of consumers and to maintain and promote effective competition in telecommunication activities. For a further discussion of the Director General's powers and duties see below.[8]

5. Licensing

The Act rendered the running of telecommunication systems and the provision of telecommunication services over them subject to a complex and sometimes rigorous licensing regime. Under this regime the policing of telecommunication licences is one of the prime functions vested in the Director General. Before 1984 licences such as that issued to Mercury did not in terms define the role of the licensee, but the Act created a new class of potential licensee, to be designated under section 9 as a "public telecommunications operator." Such operators, the prime example being BT, would (pursuant to section 8) not only have to accept certain customer service obligations in the running of their system but would also be vested with extensive powers enabling them to install apparatus over both public and private land. These powers are set out in the Telecommunications Code contained in Schedule 2 to the Act.[9]

As regards equipment, the Act put the now liberalised supply of equipment under the control of the Secretary of State who has in fact now authorised the Director General to exercise these powers pursuant to section 22(1) of the Act. The complexity of approval procedures to date has caused a large number of criticisms from equipment suppliers and the assumption of responsibilities by the Director General was perhaps a move designed to streamline such procedures, the full effects of which have yet to be seen.

OFTEL now has a staff of over 120 and the Director General must be subject to an increasing call on his time by the various pressure points within the telecommunications industry. His success or failure will most certainly be judged by his ability to promote the interests of consumers, to enhance service competition and judiciously to limit BT's apparent capability for indulging in some monopolistic activities without at the same time undermining its normal commercial flexibility and possibly weakening its strength internationally. The dilemmas this can pose for the regulators and their officers have already been seen in the recent report by the Monopolies and Mergers Commission on the proposed merger, now completed, between BT and Mitel Corporation.[10]

C. THE FUTURE: DOMESTIC AND INTERNATIONAL REGULATORY TRENDS

Such is the dynamic state of evolution of telecommunications worldwide that by the time this book is published there will have been further developments both in United Kingdom and international regulation. At home the Department of Trade and Industry has set up a Steering Group to advise on prospective infrastructure developments for the industry and has commissioned a study to identify ways in which this infrastructure might be expected to develop over the next two decades. This study, in its interim report, already points towards possible further liberalising moves, such as local cable TV companies being eventually allowed to carry telephony without BT or Mercury participation, a very limited number of additional national "fixed link" PTO licences being issued and two further DBS TV channels being licensed.

In Europe the Green Paper issued by the European Commission,[11–12] whilst in many ways a relatively conservative document, includes in its "proposed positions" first of all free (unrestricted) provision of terminal equipment, and of telecommunication services other than basic service (*e.g.* voice telephony) and secondly the separation of regulatory and operational activities of telecommunications administrations (*i.e.* PTTs). The United Kingdom is well on the way to final achievement of the first of these objectives but some way off the second.

The European Commission has now announced its intention to issue, before the end of 1988, a directive on the "progressive opening up of telecommunication services from 1989 onwards and the problem of separation of operational and regulatory functions".

Notes

[1] *A-G* v. *Edison Telephone Co. of London* (1880) 6 Q.B.D. 244.
[2] Com (87) 290.
[3] For more detailed comment see Chap. 8 below at p. 125.
[4] The BSGL and VADS licence; for further discussion see at p. 83 ("Services") and Chap. 6 below at p. 96.

[5] November 17, 1983, House of Commons Official Report, Standing Committee A p. 686.
[6] see p. 78.
[7] see p. 58.
[8] Chap. 2 below at pp. 16 to 20.
[9] See also Chap. 10 below at p. 151.
[10] Cmnd. 9715 (1986).
[11–12] See n. 1 above and Com (88) 48.

CHAPTER 2

REGULATION

The United Kingdom telecommunications industry may have been liberalised but it is now probably the most heavily regulated in the World. The British Telecommunications Act 1981 was but an hors d'oeuvre for the dish that was to arrive on the table in 1984, the Telecommunications Act of that year (referred to throughout this book as "the Act") and the equally voluminous system licences issued at the same time to PTOs such as BT.

The regulatory regime introduced by the Act could not have been properly implemented without an administrative body, backed up with suitable powers of supervision and enforcement: for this the model of the United Kingdom's Office of Fair Trading was adopted, with the creation (with more extensive powers and responsibilities than the Director General of Fair Trading) of the post of Director General of Telecommunications and the establishment of his Office of Telecommunications.

A. THE DIRECTOR GENERAL OF TELECOMMUNICATIONS

1. Functions and Responsibilities

Under section 1 of the Act the Secretary of State is required to appoint the Director General, for the purpose of performing the functions assigned or transferred to that officer under the Act. As previously mentioned[1] the person so appointed was Professor B. V. Carsberg, who remains in office today.

Section 3 of the Act defines the duty of each of the Secretary of State and the Director General, which is to exercise the functions assigned or transferred to him under Part II or III of the Act "in the manner which he considers is best calculated:

"(*a*) to secure that there are provided throughout the United Kingdom, save insofar as the provision thereof is impracticable or not reasonably practicable, such telecommunication services as satisfy all reasonable demands for them including, in particular, emergency services, public call box services, directory information services; maritime services and services in rural areas; and

(*b*) without prejudice to the generality of paragraph (*a*) above, to secure that any person by whom any such services fall to be provided is able to finance the provision of those services."

The phrase "reasonable demand" appears in Conditions 1 and 5 of BT's Licence; we now have the benefit of the first pronouncement by the Director General on the meaning of this phrase, discussed in Chapter 4.[2]

The wording of Section 3, though in lofty tone, is generalised and perhaps unlikely to be of much direct comfort or assistance to those concerned with the maintenance of competition and the promotion of the interests of the consumer. Although the Director General's Section 3(1) duties should take precedence over those set out in Section 3(2), for consumers and industry perhaps a better and more focussed approach is to be found in section 3(2). This states that the Secretary of State and the Director General each has a duty to exercise his functions under the Act "in the manner which he considers is best calculated:

(*a*) to promote the interests of consumers, purchasers and other users in the United Kingdom (including in particular those who are disabled or of pensionable age) in respect of the prices charged for and the quality and variety of, telecommunications services provided and telecommunication apparatus supplied;

(*b*) to maintain and promote effective competition between persons engaged in commercial activities connected with telecommunications in the United Kingdom . . .

(*g*) to enable persons providing telecommunication services in the United Kingdom to compete effectively in the provision of such services outside the United Kingdom;

(*h*) to enable persons producing telecommunication apparatus in the United Kingdom to compete effectively in the supply of such apparatus both in and outside the United Kingdom."

The functions given to the Director General by Parts II and III of the Act include:

(a) licensing of telecommunication systems (section 5–11);
(b) modification of licence conditions (sections 12–15);
(c) enforcement of licence conditions (sections 16–19);
(d) approval of contractors and apparatus for the purposes of licences (sections 20 and 22);
(e) keeping of registers of licences, approved contractors and approved apparatus (sections 19, 21, 23);
(f) reviewing and collecting information as to all activities connected with telecommunications (section 47);
(g) publishing appropriate information and advice for consumers and other interested persons (section 48);
(h) investigating complaints about the provision of services and the supply of apparatus (section 49);
(i) exercising powers under the Fair Trading Act 1973 and the Competition Act 1980 in relation to monopoly situations and anti-competitive practices (section 50).

Functions (a) and (d) may only be exercised by the Director General with the consent of, or in accordance with a general authorisation given by, the Secretary of State. Such consent or authorisation has been given in respect of function (d), for approval of apparatus, but has not so far been given in respect of the licensing function.

2. Exercise of Powers and Administrative Law

The exercise of the Director General's powers and duties is governed by the same laws and procedures as apply generally to Government-appointed officials. If he fails to carry out any of his statutory duties he may incur civil liability for any damage thus caused. On the other hand his wrongful failure to exercise his discretion could be challenged by an order of *mandamus* to compel performance of his duty to act.

Again as with any officer of Government, a person claiming to be aggrieved by any act or omission of the Director General would be able to petition either House of Parliament on the matter. Members of the public may also address complaints to their M.P. who can in turn refer the matter to the Minister responsible, for it to be pursued by means of a parliamentary question or referred for investigation by an ad hoc enquiry or tribunal.

Lastly it should not be forgotten that complaints by persons claiming to have sustained injustice in consequence of maladministration of a Government department such as OFTEL may be forwarded by an M.P. for investigation and report by the Parliamentary Commission for Arbitration—the Ombudsman.

Judicial Review

In addition to orders of *mandamus*, the decisions or determinations of the Director General could in themselves be challenged in either of two ways, namely:

(i) prohibition, where an order is sought preventing the particular act taking place; and
(ii) certiorari, where an order is sought to quash the particular act.

However, the courts are understandably reluctant to

interfere in the decisions of executive authorities given in exercise of their discretionary powers. Essentially the only grounds upon which they are prepared to overturn such decisions are any of the following,[3] namely:

(a) illegality;
(b) irrationality; or
(c) procedural impropriety.

Illegality would essentially have to be constituted by the Director General acting *ultra vires* or failing to have regard to relevant considerations or taking into account irrelevant considerations; procedural impropriety would involve the Director General in having failed to follow specified procedures set down in the Act or in any telecommunication licence, whereas irrationality would equate with "Wednesbury unreasonabless"[4]; in other words and in general terms, when he exercises his powers the Director General must act reasonably.

To date, none of the decisions or determinations of the Director General has been challenged in any of the above mentioned ways.

3. Consumer Interests

The interests of consumers in particular were a subject of much debate during the passage of the Telecommunications Bill. Provision was made for these interests to be monitored by four Advisory Committees ("ACTS") which were established in England, Wales, Scotland and Northern Ireland in order to assist the Director General and guide him on matters of more localised concern. The Secretary of State was also empowered to appoint advisory bodies to assist the Director General on the needs of the elderly and disabled as well as those of small businesses.

The Committee advising on telecommunications for disabled and elderly people is known as DIEL and the Committee for small businesses (which are generally referred to as businesses with under 200 employees) is known as BACT.

4. Codes of practice for Consumers

Another aspect of the Director General's responsibilities of direct importance to consumers concerns the terms and conditions (not strictly under customer contracts) or standards (in a non-technical sense) according to which BT and other PTOs would provide their services. In this connection PTO licences require the PTO to consult with the Director General on a Code of Practice for consumer affairs and publish this code, which most PTOs (including BT and Mercury) have now done.

These codes cover such matters as arrangements for fault repairs, payment of bills and compliance procedures. BT's and Mercury's licences also require them to offer a small claims arbitration procedure and details of each operator's scheme are set out in its Code of Practice. The procedure is that if a customer has a claim relating to one of the services covered in the Code and the amount he is claiming is less than a fixed amount[5] (which can be determined from time to time by the Director General), or does not involve a complicated issue of law he may put the dispute to arbitration. The Chartered Institute of Arbitrators administers the schemes and appoints an arbitrator who will come to a decision on the basis of written evidence and written submissions provided by both parties.

It is not immediately clear from BT's Code whether, in a case where BT itself has a claim against one of its customers for an amount falling within the specified limit, the customer can then require that the claim be submitted to arbitration in this way. However, BT's and Mercury's standard conditions do stipulate that any dispute satisfying the criteria mentioned above, whoever may be the claimant, can be so submitted.

5. Confidentiality of Customer Information

PTO licences also require them to prepare and "take all reasonable steps to ensure that" their employees adhere to a Code of Practice in relation to the confidentiality of customer information. This Code restricts the persons to whom customer information may be disclosed and therefore provides a model for all PTOs employees to follow in their dealings with customers.

Although applicable to all PTOs, the main reason for such a Code of Practice being required was to ensure that as BT is the monopoly supplier of network services it should not have an unfair advantage over its competitors in relation to its apparatus supply business. The Code for its Systems Business therefore requires that BT's apparatus supply business should not have access to, or make use of, sales leads obtained through its systems business. BT (as well as Mercury) is also required to have a separate Code for its "Supplemental Services" (*i.e.* VADS) business, the regulatory authorization for which is discussed below.[6]

6. Complaints about Telecommunication Services

The ACTS are intended to be OFTEL's prime link with consumers, through whom their concerns and opinions are to be heard, and these Committees are also to provide a channel of communication between consumers and the suppliers of services and apparatus. Within each of the four ACT regions there have also been established local Telecommunications Advisory Committees known as "TACS." Each ACT provides an annual report which is published along with the Director General's own report for each calendar year.

7. Exceptions for National Security and Conduct detrimental to consumers

Section 3(3) states that the duties imposed upon the Secretary of State and the Director General do not apply in relation to anything done by the Secretary of State in the interests of national security or relations with the Government of a country or territory outside the United Kingdom, or in the exercise of functions assigned or transferred by or under section 50 of the Act.

Section 50(1) requires the Director General to exercise the functions of the Director General of Fair Trading under Part III of the Fair Trading Act 1973, if he is so requested by the Director General of Fair Trading. (Part III of the Fair Trading Act applies to courses of conduct detrimental to the interests of consumers.)

Section 50(2) and 50(3) respectively transfer to the Director General, to be exercisable "concurrently" with the DGFT, the functions of the DGFT under the 1973 Act with respect to monopoly situations affecting "commercial activities connected with telecommunications" and the functions of the DGFT under the 1980 Act with respect to anti–competitive courses of conduct affecting the supply of telecommunication apparatus or services.

8. Quality of Service

One of the particular responsibilities undertaken by the Director General concerns monitoring and encouraging improvements in the quality of service provided by PTOs, particularly BT. Before privatisation BT published its own information about quality of service but discontinued this practice. It has now resumed issue of quality of service statistics or indicators and Mercury intends to do the same.

PTO licences do not contain specific quality standards and requirements. It appears to have been assumed that quite apart from the impact of competition in enhancing service quality BT as an established operator and Mercury, with a new digital system, will provide acceptable quality although when their systems are interconnected this will not necessarily be the case so long as BT has a substantially analogue system. Accordingly in relation to interconnection of BT and Mercury's systems specific provision as to quality of service was made in the Director General's determination of October 11, 1985.[7]

The Director General stated, in his November 1986 report on quality that he regards the obligation to provide universal service (Condition 1 of BT's licence) as implying that service must be of satisfactory quality. The point arose in the early part of 1987 when, due to strikes and equipment problems, BT's service worsened. At the time of writing standards have improved but it is difficult to see that any particular standard of quality (other than call quality which is so poor it is tantamount to no service at all) can be extracted from the wording of Condition 1 of BT's licence.

The particular factors taken into account by the Director General in looking at quality of service include waiting times for telephone installations, call failure rates, directory enquiries and operator services, faults and fault repair services, public call boxes and the public perception generally of the level of service provided. However, for the time being he has stated he will rely more on BT's own statistics and carry out his own independent investigations in order to monitor not only BT's performance but BT's own assessment of that performance.

9. Price Control

This is one of the more sensitive if not the most sensitive area of regulation subject to the Director General's supervision and a rock on which he must light a large beacon for himself.

BT's licence at present contains a specific rule to control changes in the prices of certain core services, namely business and domestic exchange line rentals and direct dialled inland telephone calls except those made from public call boxes. The rule is that a weighted average of price changes for these services is limited to three percentage points below the rate of inflation, known as "the RPI–3 Rule."

Currently BT introduces price changes in November of each year. The first changes in the Director General's period of tenure, in November 1985 raised a chorus of complaint and the changes in November 1986 provoked a similar barrage. However the Director General was well prepared: in his Review of the same month he demonstrated why BT had complied with its licence obligation notwithstanding that prices of certain services had risen by more than the licence formula since these were offset by net falls in certain tariffs. (In August 1987, BT announced that it would not then be increasing its prices for local and national calls and exchange line rentals.)

Condition 24 of BT's licence provides that the price

control should cease to apply after the period expiring July 31, 1989 when simple resale may be allowed.[8] However, the Director General looks set to seek modifications to BT's licence in order to continue the same or a modified form of price control even after this date.[9] (For further discussion of licence modification procedures see below.)[10] If BT were not prepared to agree to this the Director General would have to refer the matter to the MMC. It should be noted that in the meantime the Board of BT have also given a voluntary undertaking to limit increases in the rentals of residential exchange lines to a level equivalent to the product of RPI–2.

The Director General is also empowered to investigate other pricing policies of BT in relation to services not covered by the RPI–3 Rule and has for example done so in relation to the prices of access lines and private leased circuits.[11] (He has not as yet taken any action pursuant to his investigation of these matters.) However any regulatory controls over such prices are to be found only in the more general fair trading conditions embodied in BT's licence (discussed below).[12]

10. Promotion of Competition

The Director General is required by the Act to promote competition as a means essentially of bringing the greatest benefit to consumers. He therefore attaches a high priority to his duty to promote effective competition and, in view of BT's monopoly position, the Director General has stated that he proposes actively to seek out examples of breaches of licence conditions. Presumably in a move to try to ensure due compliance without his intervention, at an early stage he encouraged BT to produce yet another Code of Practice entitled "Competitive Marketing Guidelines." This particular bible was directed to BT's employees as a way of stressing the importance of fair dealing in their relations with customers and their competitors. The code does not have any status under BT's licence but does at least reproduce elements of BT's own licence obligations.

The policing of competition in telecommunications covers an extremely wide spectrum in which the Director General will occasionally find he is not ideally supported by the regulatory powers with which he has been provided. In telecommunications services, where BT's monopoly power is at its strongest, the Director General is required to oversee the position of:

(i) customers,
(ii) competitive providers of basic services (*e.g.* Mercury),
(iii) competitive providers of value added services, and
(iv) providers of other services utilising the BT system, for example telephone information service providers.

Moreover because the regulatory system is primarily not capable of direct and individual enforcement, for example in the Courts, this must have the effect of funnelling all claims and complaints to the Director General himself. As a consequence his staff will over time come under increasing pressure and, given that there is only one person who can ultimately make necessary decisions of importance, (even though perhaps on recommendations from the Deputy Director General or others in his staff), it does seem unlikely that all complainants will be able to receive equal treatment and the fullest or most prompt attention in all cases. Priorities are almost bound to come into the matter.

The most direct method of enforcement of competition by the Director General will be through his powers to issue licence compliance orders under sections 16 to 18 of the Act. This procedure and its effectiveness is discussed below.[13]

In his "watchdog" role the Director General has already received complaints in a number of areas, particularly relating to BT's own licence. The first major matter concerned radio paging where the complaint was made by BT's then four competitors in the provision of radio paging services. The complaint was that BT's practice of joint billing for radio paging services and basic telephone services and its method of charging for radio paging services gave it an unfair advantage and therefore inhibited effective competition. In his report of August 1985 the Director General concluded that BT should be required to introduce separate billing for its radio paging customers as soon as practicable and subsequently BT agreed to a modification of its radio paging licence (a separate licence from its licence as a PTO) establishing this practice as a licence obligation. The Director General was not prepared to go further as requested by the complainants and modify Condition 17 of BT's PTO licence so that any unfair favouring of one of its businesses in relation to any activity would have been prohibited.

In 1985 the Director General was also called upon to deal with complaints against BT allegedly leaking confidential customer information and cross-subsidising in relation to BT's security services. These services involved monitoring the state of a telephone circuit and the transmission of an alarm to emergency services. Security companies competing with BT claimed that BT had circulated marketing literature to customers whose names would have been obtained in the course of work by other parts of BT providing basic telephone services to such customers at the behest of the security companies. On the basis of BT's code of confidentiality of customer information being in force at the relevant time it was alleged that this activity by BT would have been in breach of the code.

BT's licence obligations as to confidentiality are briefly explained later in this book.[14] In this case BT argued that its own security service was a basic service rather than a value added service and that therefore the rules in its licence on cross-subsidisation and disclosure of confidential information did not extend to it. However the Director General appears to have persuaded BT otherwise and BT voluntarily agreed to observe its licence rules on fair trading in relation to such business.

11. BT and Mercury

Effective competition between BT and its only licensed competitor in the provision of basic services, Mercury, was always likely to be an area where the Director General's actions came under close scrutiny. Litigation has already ensued between both companies and the Director General has twice been called upon to intervene pursuant to Condition 13 of BT's licence in order to determine the necessary terms and conditions for an interconnection agreement between the two parties, in relation first to telephony (October 1985) and subsequently in relation to international telex (July 1987). He has also made a separate determination for them in relation to international accounting. These determinations are discussed below.[15]

12. Value Added and other Service Providers

As regards other services, the amount of policing activity in which the Director General is likely personally to be involved, for example in relation to value added services, is difficult to gauge at this early stage. However, given the potential number of VADS operators in the market the licence constraints on the scope of services authorised appear likely to be tested quite frequently as may the enforceability of fair trading conditions applicable to both BT, Mercury and major service providers (see Conditions 1(m) and 9 of the VADS licence). For further discussion of VADS see below.[16]

B. Wireless Telegraphy

The Wireless Telegraphy Act 1949 defines (section 19) "wireless telegraphy" as meaning:

> "the emitting or receiving, over paths which are not provided by any material substance constructed or arranged for that purpose, of electro-magnetic energy of a frequency not exceeding 3 million megacycles a second, being energy which either—
>
> (*a*) serves for the conveying of messages, sound or

visual images (whether the messages, sound or images are actually received by any person or not), or for the actuation or control of machinery or apparatus; or

(*b*) is used in connection with the determination of position, bearing or distance, or for the gaining of information as to the presence, absence, position or motion of any object or of any objects of any class."

The principal effect of the WTA is to make it an offence for anybody to establish or use any "station for wireless telegraphy" or to install or use any apparatus for wireless telegraphy without a licence for that purpose granted by the Secretary of State. There are certain exemptions from the licence requirement, for example with respect to broadcast relay apparatus, such as SMATV.

From the point of view simply of telecommunication, the important thing to note is that as in the case of telecommunication systems, wireless telegraphy includes not only emission but also receipt of messages. Accordingly the licence requirements extend to receiving stations as well as transmitters and thus include mobile telephones and radio paging devices. All such apparatus would normally be included in the same wireless telegraphy licence granted to the operator of the particular service.

Notwithstanding the exemption for broadcast relay apparatus, the apparatus used by subscribers to the services of broadcast relay companies, as well as of cable television operators, is also deemed to be wireless telegraphy apparatus by virtue of the proviso to section 19(1) of the WTA. The Secretary of State has issued a general telecommunications licence for such apparatus under section 7 of the Act.[17]

Wireless telegraphy licences are quite separate from licences required under the Act. Since wireless telegraphy is a means of telecommunication, any station or apparatus for wireless telegraphy would also represent a telecommunication system which, unless it falls within one of the exceptions in section 6 of the Act, would require a licence under section 7 of the Act.

A wireless telegraphy licence may be revoked by the Secretary of State at any time pursuant to his power under section 1(4) of the WTA; according to the DTI he would exercise this power if the apparatus comprised in the wireless telegraphy station or its performance at any time became "unacceptable" to him. In relation to PTOs this power has been curtailed, in order that licences granted to an operator under both the Act and the WTA may only be subject to revocation in similar circumstances.[18]

Notes

[1] At p. 12 above.
[2] At p. 49 below.
[3] *Per* Lord Diplock in *Council of Civil Service Unions* v. *Minister for the Civil Service* [1984] 3 All E.R. 935 at 950.
[4] *Associated Provincial Picture Houses Ltd.* v. *Wednesbury Corporation* [1947] 2 All E.R. 680. See also *Preston* v. *IRC* 1985 2 All E.R. 327 & *Wheeler* v. *Leicester City Council* 1985 2 All E.R. 1106.
[5] Currently £1000; see also p. 76 below.
[6] See Chap. 4 below at p. 77.
[7] See Chap. 4 below at pp. 67 to 70.
[8] See further Chap. 4 below at p. 78.
[9] See "The Regulation of BT's Prices: A Consultative Document" OFTEL, January 1988 and see Chap. 4 below.
[10] See Chap. 3 at p. 38.
[11] Prices of Access Lines and Private Circuits: A Consultative Document OFTEL 9/86.
[12] See Chap. 4 at pp. 59 *et seq*.
[13] See Chap. 9 at p. 138.
[14] See Chap. 4 at p. 75.
[15] *Ibid.* at pp. 55 and 67.
[16] Chap. 6.
[17] Class Licence for Telecommunication Apparatus for reception of services conveyed by means of certain cabled systems November 25, 1986.
[18] S. 74 of the Act.

CHAPTER 3

LICENSING OF TELECOMMUNICATION SYSTEMS: OVERVIEW

1. The Licence Requirement: Prohibition and Authorisation

Fundamental to the United Kingdom regulatory regime for telecommunications is the requirement for the licensing, on an individual or class or general basis, of any telecommunication system. The licensing provisions are set out in Part II of the Act.

At the outset it should be noted by anyone studying a telecommunication licence for the first time that the structure of licences is particularly important. These follow a standard format, with the licence grant taking up a few paragraphs on the first page and incorporating by reference to Annex A a description of the systems licensed and, by reference to Schedule 3, the acts (including services) which the licensee is authorised to do in relation to those systems; the conditions to which the licence is subject are set out in Schedule 1 and the Secretary of State's powers of revocation of the licence are set out in Schedule 2.

Generally the official approach on the formulation of telecommunication licences is that the licensee is able to use the system to do anything which is not expressly prohibited, as opposed to doing only those things which are expressly permitted.

Section 5 provides that (subject to certain limited exceptions—see below) a person who runs a telecommunication system within the United Kingdom shall be guilty of an offence unless he is authorised to run the system by a licence granted under section 7.

Even though a system is properly authorised in this way, the applicable licence will not prevent the commission of an offence by the person running the system (variously referred to as the "licensee" or the "operator") if another system or apparatus is connected to the licensed system or services are provided by means of that system which in either case are not authorised by the licence to be so connected or provided (section 5(3)).

A person guilty of any of these offences is liable:

(a) on summary conviction, to a fine not exceeding the statutory maximum; and
(b) on conviction on indictment, to a fine.

Currently the statutory maximum[1] is £2000 and the maximum fine on indictment is set according to the offender's capacity to pay. It is a defence under section 5 for the person charged with the offence to prove that he took all reasonable steps and exercised all due diligence to avoid committing the offence.

Since the unauthorised running of a telecommunication system can result in the commission of an offence, it is important to understand what constitutes a telecommunication system. The word "system" is not defined either in the Act or in any of the licences issued under it, but "telecommunication system" is defined as follows, in Section 4(1) of the Act:

> ". . . a system for the conveyance, through the agency of electric, magnetic, electro-magnetic, electro-chemical or electro-mechanical energy, of—
>
> (*a*) speech, music and other sounds;
> (*b*) visual images;
> (*c*) signals serving for the impartation (whether as between persons and persons, things and things or persons and things) of any matter otherwise than in the form of sounds or visual images; or
> (*d*) signals serving for the actuation or control of machinery or apparatus."

A message could be conveyed by all kinds of means but as

can be seen a distinguishing characteristic of a telecommunication system under the Act is that its messages require an input of the type of energy specified in section 4(1). In a typical case it thus seems probable that it would be the person who controls the apparatus permitting such energy to enter the system who would be held to be running that system. Moreover the Act itself tacitly endorses this approach in one set of circumstances referred to in section 4(2) of the Act.

Although not a general authority on the point, this Section provides:

> "For the purposes of this Act telecommunication apparatus which is situated in the United Kingdom and—
>
> (*a*) is connected to but not comprised in a telecommunication system; or
> (*b*) is connected to and comprised in a telecommunication system which extends beyond the United Kingdom,
>
> shall be regarded as a telecommunication system and *any person who controls the apparatus shall be regarded as running the system.*" (Emphasis added.)

In general, therefore it seems fair to assume that any person controlling the apparatus comprised in a telecommunication system is running that system but in any case where the question is crucial, an appraisal of all relevant factors, including not only the person controlling the apparatus, but also its ownership, the premises where it is located and the person sending messages by means of the apparatus should be carefully made.

The specific reason for the inclusion of section 4(2) was primarily to prevent circumvention of the licensing requirement by the use of apparatus, such as a satellite ground station, connected purely internationally, for example by wireless telegraphy, to an orbital satellite. The system in which such apparatus would be comprised would extend beyond the United Kingdom and, but for the wording of Section 4(2), would have fallen outside section 5.

As is reasonably evident from the above, even the humble telephone (*i.e.* the instrument itself), certainly with its associated wiring, is a telecommunication system requiring a licence under the Act. This would normally be the BSGL.[2]

Each licence granted under the Act has its own integrity: by this I mean that the licensee is required to comply with all its terms and conditions if it is to apply to him and cannot rely in part on one licence and in part on another in order to validate the running of a particular system. Only one licence (class or individual) can therefore be applied to that system at any one time.

Confusion sometimes arises over the difference between running a network and running a system. It is important to make a clear distinction: telecommunication regulation is only concerned with the running of systems; a network may comprise different systems run by different persons, for example nodes operated and therefore run by a data service provider linked together by private circuits run by PTOs. Similarly, a system may be run by one person but the services provided over it may be offered by another, perhaps associated, entity. In these circumstances compliance with the licence is a responsibility of the person running the system, who must make it his business to ensure that the provider of the services does not cause him to be in breach of the licence.

However, in such circumstances both the licensee and the service provider could be guilty of an offence; section 5(4) of the Act provides that where a person commits the offence of running an unlicensed system because of the default of some other person, that other person is also guilty of an offence.

Provision of unlicensed services

Where the services to be provided by the operator of a telecommunication system are not covered by a licence granted under the Act, by virtue of section 5(2) any contract for their provision in this way is liable to be illegal and therefore unenforceable by either party.[3]

2. Limited Exceptions

The general prohibition on the running of unlicensed systems does not apply to systems run by broadcasting authorities where the transmissions are either by wireless telegraphy for general reception or closed circuit television on a single set of premises. The prohibition is also not contravened by a system which simply involves the conveyance of light readable by the human eye, nor by the running of a system confined to one set of premises or in a vehicle, vessel, aircraft or hovercraft. There is a further exception covering private systems, such as baby transmitter/receivers. This is in terms of the running by a single individual of a telecommunication system unconnected to any other system and whose apparatus is under the control of that individual and where the system is run purely for domestic purposes (section 6 of the Act).

Such is the breadth of the definition in section 4(1), were it not for the fact that a system appears to have to comprise apparatus (telecommunications apparatus being itself defined in section 4(3) of the Act) a teasing, albeit somewhat academic point, could perhaps have been made. That is whether the human brain and its connected organs constitute a telecommunication system. However, there is no doubt their unlicensed use would anyway have been excused by section 6!

Perhaps the most important exception, at least for business, is in sections 6(3) and (4) which contain important exemptions where the sounds or visual images conveyed by the system (which must not be connected to any other system) cannot be heard or seen by anyone other than the person carrying on the business or his employees and certain other conditions are satisfied, *e.g.* no services are provided to others by means of the system. This would cover not only "in–house" systems but also privately provided systems[4] as well as, for example, closed circuit television systems.

3. Granting of Licences

A telecommunication licence may be granted by the Secretary of State, after consultation with the Director General, or by the Director General himself provided the Director General is so authorised by the Secretary of State (which authority has not yet been given). Accordingly the Value Added and Data Services Licence,[5] (a class licence) and the Branch Systems General Licence,[6] (another class licence) were issued by the Secretary of State.

The Act confers on its administrators rather greater flexibility in the administration of individual, as opposed to class, licences, for two reasons. First, section 7(6) provides that individual licences may require the licensee to comply with directions given by the Director General or prohibit the licensee from doing something unless the Director General consents otherwise; such individual licences may also provide for the reference to the Director General for his determination of questions arising under the licence; the interconnection condition in PTO licences (see Chapter 4 and Condition 13 of BT's Licence) is the prime example.[7] It is arguable that by virtue of section 7(6) such discretionary powers cannot be included in class licences.

Secondly, although the conditions (but not the provisions of other schedules or in the Annexes) of a licence may be modified, in the case of an individual licence, with the consent of the licensee (and even without its consent)[8] in the case of a class licence no modification can be made unless no representations or objections whatsoever are made by persons covered by the licence or any such representations or objections are withdrawn. Accordingly, a modification of a class licence in this way is unlikely to be practicable and is more likely to be accomplished by its revocation and re-issue in the modified form. Alternatively, the Director General could refer a class licence modification to the MMC, under Section 13 of the Act.[9]

To date, individual licences have been the prerogative in practice of the Secretary of State and his officials at the Department of Trade and Industry. However, in practice application for an individual licence would usually have to be made to OFTEL, with whom questions arising on the application would be pursued. The grant of such a licence can take many months, involving as it does an investigation of the nature of the particular system, the reasons why a class licence is not applicable, whether any policy considerations are involved in relaxation of any restrictions normally applied on the running of the system and in many cases a comparison with other licences granted to other individuals or classes in similar circumstances.

4. Enforcement of Licence Conditions

The structure of telecommunication licences and sections into which they are divided are liable to cause confusion, at least to newcomers, in that different sections have different legal implications. As already pointed out (section 1 above), it is an offence to run a telecommunication system that is unlicensed; thus any provision contained in the licence grant itself (usually one or two pages at the very beginning of the licence), and in its Annex A and Schedule 3 must by virtue of sections 5(1) and 5(2) of the Act, be observed in order to avoid commission of a criminal offence. However, except as mentioned below, a breach of a licence condition (other than non-payment of a licence fee) does not trigger any immediate legal sanction in that for example this cannot *per se* be punished by revocation of the licence. The procedure is in fact that the Director General may issue a licence compliance order[10] and if that order is not complied with the Director General may revoke the licence or request the Court to grant a mandatory injunction requiring such compliance, failing which the licensee would be in contempt of court and the court could punish the offender with a fine or imprisonment. Under section

18(6) of the Act such a failure to comply with the Director General's order would constitute a breach of duty for which the licensee would be liable to any person who suffers loss or damage as a result of the breach.

Only if a licensee fails to comply with a final order (under section 16 of the Act) or a provisional order confirmed under that section (where such provisional order is not subject to review) and such failure is not rectified within three months after the Secretary of State has given notice in writing of such failure to the licensee is the licence finally revocable (on 30 days notice in writing to an individual licensee or any licensee covered by a class licence).

Breach of a licence condition can have immediate legal consequence in one case. BT and other PTOs' standard conditions of contract contain terms entitling the PTO to terminate service immediately in the event of the customer contravening his licence (typically, for the generality of customers, the BSGL). This right, for example in BT's case, as with other PTOs, is reinforced by Condition 53.6 of its licence which excuses BT from its obligation to provide service where the customer is in breach of contract.

5. Modification of Licences

Conditions (*i.e.* those usually contained in Schedule 1) of licences granted under section 7 of the Act may be modified essentially in two ways, either by agreement or by reference to the MMC. As pointed out above, however, class licences are not readily susceptible of modification by agreement, given the representation and objection procedures which must be followed under section 12(4) of the Act.

As regards licence modifications by agreement, the procedure set out in section 12 of the Act is that the Director General should first publish a notice stating the proposal and the reasons for the modification and specifying a time, of not less than 28 days, within which representations or objections may be made. A copy of such notice is also to be sent to the Secretary of State and if within the specified time the Secretary of State directs the Director General not to make any such modification, the Director General must withdraw the proposal. Section 12(6) sets out the criteria which govern any such direction by the Secretary of State.

Licence modification references to the MMC arise where the Director General has been unable to make the modification by agreement with the licensee. Such references may also be made where the Secretary of State has given a direction under section 12(6)(*a*) of the Act or where the Director General himself considers the matter of sufficient importance to refer direct to the MMC. The procedure for a MMC reference is set out in section 13 of the Act. Under its provisions the reference is to be framed to ask the MMC to investigate and report on whether the particular circumstances giving rise to the proposed modification operate, or may be expected to operate, against the public interest and whether these circumstances, "the effects adverse to the public interest," could be remedied or prevented by the proposed modifications.

As with an agreed modification, the Secretary of State can intervene to direct the MMC not to proceed with the reference (section 13(5)).

Following the MMC Report, which must be provided within six months from the date of the reference,[11] where this concludes that the matters specified in the reference operate, or may be expected to operate, against the public interest, and specifies appropriate modifications to deal with the problem, the Director General must make the necessary modifications. He must however first give the same notice as is required in relation to a licence modification by agreement and then consider any representations or objections so made. Again the Secretary of State may then direct the Director General not to make modification, where it appears to him that this is requisite or expedient in the interests of national security or foreign relations.

As discussed below,[12] section 95 of the Act also empowers the Secretary of State to make licence modifications following a monopoly, merger or competition reference.

6. Fees for and Publication of Licences

A licence fee is payable on the grant of the licence, the amount of which is intended to contribute towards the costs of administration of the licence and as such it tends to be a reasonable sum. Currently, the DTI's policy tends to be to charge about £2000 (or £100 for a temporary licence) for the grant of individual licences for systems not involving the provision of services to others. PTO licensees are required to pay a great deal more, namely a lump sum down payment and regular annual amounts to cover the Director General's and associated costs, subject to a maximum; in BT's case this maximum is 0.08 per cent. of the annual turnover of BT's System Business. Every Major Service Provider who is a licensee under the VADS class licence is to pay an amount determined from time to time by the Director General. At present this amount stands at £1,000.

All licences granted under section 7 are published and OFTEL publishes a monthly Publications List giving details of licences granted to PTOs, class licences and licences granted to particular persons, copies of which are available on payment of the appropriate fee.

7. Special Licences

Privately Provided Systems

There are a number of utilities, for example British Gas Corporation, which have systems for transporting their own products to consumers (*e.g.* gas pipelines) but which are obviously highly susceptible of adaptation for telecommunication (*e.g.* electricity power lines) or for physical combination alongside telecommunication links (*e.g.* cables). Similarly organisations like the British Railways Board require signalling systems linked together by cables and wires which can also serve as telecommunication systems for other purposes, *e.g.* voice. Armed with powers as statutory undertakers as well as the necessary telecommunication licence rights many of these organisations are thus able to install their own links without reliance upon a PTO.

Accordingly, over the years these utilities have built up their own telecommunication systems and provided their own links for this purpose. Under current rules this activity is perfectly lawful and does not require to be licensed so long as the system is not connected to any other telecommunication system. In fact many utilities' systems must have such connections, particularly to the public networks of BT and Mercury, as a number of them have licences under the Act, including British Gas Corporation, the Central Electricity Generating Board and the British Railways Board. These connections will be for such purposes as emergency voice services.

Many of these licences restrict the use of privately provided telecommunication systems for the conveyance of messages other than those relating to the business of the licensee.

Systems requiring a licence which have been privately installed may also be able to be run under the BSGL, provided the constraints in its Annex A (particularly as to distance between different sets of premises) can be observed.

BT (and Post Office) Granted Licences

One of the other licensing quirks is the existence of a number of licences granted by BT under the 1981 Act and previous legislation, which were maintained in force by virtue of paragraph 1 of Schedule 5 to the Act. Such licences are deemed to have been granted under section 7 of the Act; they cease to be valid on August 4, 1989.[13]

Although BT was required[14] to provide the DTI with documentary evidence of such licences the identity of these organisations operating under them has not been made public.

Systems First Run Before August 1984

There is also a Class Licence "for the running of certain telecommunication systems first run before 5

August 1984" granted by the Secretary of State on August 3, 1984. This licence looks like a "mopping-up" provision to cover systems such as those run with the consent of BT prior to August 1984 but where there is no written evidence of such a licence (individual licences granted by BT under the 1981 Act did not have to be in writing). As the title demonstrates, the licence can authorise the running of any system (other than one run by BT, Mercury or Hull) first run before introduction of the new licensing regulations under the Act. However, the licence can only apply to systems not otherwise licensed.

This licence, which remains in force until August 1991 is in substantially the same form as the BSGL with certain notable exceptions. The private circuit connections and rules on "hopping" do not apply to any connection and its use which was permitted by BT before August 1984. There is also a restriction on any privately provided circuit connected to the licensed system being used to convey messages to or from a public switched network, except with the consent of the relevant PTO.

8. Public Telecommunication Systems

Under section 9 of the Act, the Secretary of State may by order designate as a public telecommunication system any system the running of which is authorised by a licence granted to a particular person and which includes conditions specified in section 8 of the Act.

To date, the persons so designated as operators of public telecommunication systems, "public telecommunications operators," are British Telecom, Mercury Communications Limited, Kingston-upon-Hull City Council/Kingston Communications (Hull) PLC, the two cellular radio operators, Racal Vodafone and Telecom Securicor, and the various cable television operators. (A complete list is set out in Appendix B.[15]) The procedure for designation as a public telecommunications operator is set out in section 9. These operators are generally referred to as "PTOs" and this abbreviation is used in this book.

Generally (see the exception in 7 above) only PTOs

have the right to install and run the "fixed" links (in excess of 200 metres) between telecommunication systems. Similarly, because of the restrictions in the BSGL, only PTO's may connect different sets of "served premises," or connect different parts of the same branch system where these are more than 200 metres apart. These connections, commonly referred to as "fixed links" may take the form not only of wires and cables but also wireless telegraphy stations, aerials and repeaters.

The Telecommunications Code contained in Schedule 2 to the Act, which contains powers very similar to those previously granted to BT and the Post Office as statutory undertakers, also applies to public telecommunication operators and, exceptionally, to persons running systems where the Secretary of State has concluded that the particular system will benefit the public and that it is not practicable for the system to be run without the application of the Telecommunications Code (section 10(2) of the Act). So far as the author is aware, the only non–PTO on whom Code powers have been conferred are the Water Authorities. For a further discussion of this Code see Chapter 10.

When the Secretary of State proposes to apply the Telecommunications Code, he must publish a notice indicating his intention and the effect of any conditions he would propose to include in the licence for the purpose of safeguarding the environment, limiting damage to streets and reducing interference with traffic. Once any representations have been received and considered, the Secretary of State must publish a further notice indicating the outcome.

The conditions embodied in PTO licences are as important to customers and providers of telecommunication services as to the licensee operators themselves. Positive obligations are imposed on the operators for the benefit of customers and other operators. The licence granted to BT is, whilst in basically similar terms to those granted to Mercury and Hull, the most extensive licence granted under the Act and the best reference point for a general discussion of the regulations to which all PTOs are subject. BT's licence is therefore considered in detail in the following Chapter.

Notes

[1] See s. 106(2) of the Act.
[2] See Chap. 5 below.
[3] See inter alia *Levy* v. *Yates* 1838 8 A. & E. 129; *Dungate* v. *Lee* 1969 1 Ch. 545; *Archbolds (Freightage) Limited.* v. *S. Spanglett Limited* [1961] 1 Q.B. 374; *Allan (Merchandising) Limited* v. *Cloke* [1963] 2 Q.B. 340.
[4] At p. 40 below.
[5] Discussed at Chap. 6 below at p. 96.
[6] Discussed at Chap. 5 below at p. 80.
[7] Chap. 4, p. 66 below.
[8] See p. 38 below.
[9] *Ibid.*
[10] See further in Chap. 9 at p. 138.
[11] S. 13(9) of the Act.
[12] Chap. 9 at p. 141.
[13] S.I. 1986 No. 1275.
[14] Para. 1(2), Sched. 5 of the Act.
[15] See p. 197 below.

CHAPTER 4

THE LICENCE OF BRITISH TELECOMMUNICATIONS

As mentioned at the end of Chapter 3, BT's licence is marginally the most extensive of the PTO licences and is therefore the best example on which to comment generally on the regulatory controls applicable to PTOs.

It should be remembered that in the main the licences of BT and Mercury and indeed of Hull are virtually identical in all respects, the only substantial difference being found in Condition 1 of each of their licences. For example, whereas BT has a universal service obligation, Mercury has obligations to expand its network to certain areas by certain dates (it has in fact gone much further than these obligations in terms of the extent of its network and its accelerated installation).

BT's Licence came into force on August 5, 1984, the day before BT became a public limited company and a few months before its flotation the same year. The Licence is divided into six parts:

(i) The right to run certain telecommunication systems;
(ii) conditions regulating how the systems are to be run;
(iii) arrangements for revoking the Licence;
(iv) authorisation to connect to other systems and to provide telecommunication services;
(v) exemptions and conditions relating to the application of the Telecommunications Code; and
(vi) a description of the telecommunication systems BT is allowed to run.

This manner of division of the Licence does not particularly lend itself to the easiest explanation and so I propose to begin with a discussion of the licence grant itself and the way in which it could be terminated, to go on to the scope of the authority which it confers in the running of the licensed systems and their connection to other systems, then to consideration of the all-important conditions (1 to 17 inclusive) set out in Part 2 of Schedule 1, applied pursuant to section 8 of the Act, as well as the conditions (18 to 53) included in Part 3 pursuant to section 7 of the Act. The Telecommunications Code provisions are discussed in detail in Chapter 10.

A. THE LICENCE GRANT

This gives BT permission to run the telecommunication systems (known as the "Applicable Systems") described in Annex A to the Licence and to connect to other telecommunication systems and to provide the telecommunication services as specified in Schedule 3. In running the Applicable Systems, BT must abide by the conditions set out in Schedule 1 and the permission can be revoked, or taken away, in the circumstances described in Schedule 2, which, as previously explained,[1] do not include mere breach of a licence condition.

The BT Licence is for a minimum duration of 25 years from June 22, 1984, subject to revocation on at least ten years' notice, which may not be given before the end of the fifteenth year.

Termination of the Licence prior to the end of this 25 year period can only arise in the particular circumstances set out in Schedule 2 which include, *inter alia*, failure of BT to comply with a final Order or a provisional Order made under section 16 of the Act (where the Order is not subject to proceedings for review and the failure has not been rectified within three months of notice from the Secretary of State), or its insolvency, receivership or liquidation: all contingencies which, it might be said, seem more theoretical than real.

B. LICENSED SYSTEMS

The systems which BT is so licensed to run, set out in Annex A, demonstrate the network boundary concept underlying the Act and its regulations. A licensed system is described as one by means of which messages are conveyed or are to be conveyed from one network termination point ("NTP") to another such point, for example a normal inland call, or from an NTP to a place which is not an NTP, for example an international call or a call between other places "but in any case not beyond a Network Termination Point." The apparatus comprised in the licensed system must not include terminal apparatus installed in premises occupied by a person to whom BT provides services. This apparatus would form part of the customer's system, typically licensed under the BSGL. Also, the BT system must not be a wireless telegraphy system except where such telegraphy is provided to or from permanent or temporary fixed stations (for example microwave), thus precluding the provision of mobile services (paging, radiotelephone, etc.) under this particular licence.

C. CONNECTION OF OTHER SYSTEMS: PROVISION OF SERVICES

The other systems and apparatus with which BT's licensed "Applicable Systems" may connect are described in paragraph 1 of Schedule 3 and essentially cover licensed systems and approved telecommunication apparatus.

The services permitted by the Licence are set out in Schedule 3 paragraph 1(b) and comprise:

> "the provision by means of the Applicable Systems of telecommunication services consisting in:
>
> (i) the conveyance (not including switching) of messages (not including cable programme services sent under a licence granted under [section 58 of the Act[2]]) and switching incidental to such conveyance; and
>
> (ii) directory information services
>
> but not any Land Mobile Radio Service."

According to section 4(7) of the Act the word "convey"

and derived expressions include "transmit, switch and receive." For the purposes of Schedule 3 therefore "conveyance" should be construed in this sense.

The Explanatory Notes to the Licence once stated that value added services may not be provided under the Licence and that BT would will require a separate licence for this. This interpretation of Schedule 3 has been revised in the statement of the Secretary of State on February 25, 1987. The modifications made to conditions of PTO licences, including BT's, on May 1, 1987 reflect this revision, by referring in a number of places to the "Supplemental Services Business," itself defined essentially as the provision of a "Relevant Service": this unhelpful nomenclature is simply a rather elliptical description of the types of service which a licensee can provide under the VADS Class Licence. A PTO can offer the same services under its licence.

D. CONDITIONS

Rather than wade through and analyse each condition in turn, I believe it is more helpful to the reader to concentrate on the main substantive elements of BT's licence obligations and deal with these in what I hope are logical and coherent sections.

1. Universal Service: Conditions 1 and 46

Condition 1

Condition 1 obliges BT to provide what is ambitiously, but commonly, described as a "universal" telecommunication service. Universal, that is, in the limited national sense of throughout BT's "Licensed Area," which is the United Kingdom excluding Hull. In particular, BT must provide to every person who so requests both:

> "voice telephony services; and other telecommunication services consisting in the conveyance of messages . . . except to the extent that the Director is satisfied that any reasonable demand is or is to be

> met by other means and that accordingly it would not be reasonable in the circumstances to require [BT] to provide the services requested."

There may be other circumstances in which BT could be excused performance of this obligation—see Condition 53.[3]

Condition 1, in referring exceptionally to "reasonable demand" in this way provides an interesting contrast to Section 3 of the Act which itself requires the Director General to ensure that telecommunication services are provided in order to meet "all reasonable demands for them". Note that BT's obligation in Condition 1 is not to provide service in accordance with a reasonable demand but to provide service in whatever circumstances "except to the extent that the Director is satisfied that any reasonable demand is . . . to be met by other means". Accordingly, aside from the exceptions in Conditions 53, BT can only escape liability to provide service under Condition 1 where the demand can be met by other means.

The first occasion on which the Director General has been called upon to interpret and apply Condition 1 has just arisen in the PanAmSat case.[4] There, the would-be independent satellite operator, PanAmSat, claimed that BT was in breach of its licence Conditions 1, 5, 17 and 35 in refusing to agree to provide service linking customers in the United Kingdom to PanAmSat's satellite. The Director General gave his view that "reasonable demand is primarily demonstrated in the market: reasonable demand exists if one or more customers will pay a fair price for the service". This rather skirts around the point that the circumstances surrounding the demand may lead to, for example, the attempted application by BT of additional terms and conditions which the customers might be unwilling to accept.

Generally, therefore, as a rule of thumb, it would in the author's view be safer to assume "reasonable demand" as meaning a demand which in all the circumstances and, after taking account of the exceptions in Condition 53, is reasonable. The matter then becomes in effect a jury question, one for the Director General to interpret exercising his discretion reasonably.

Condition 1 goes on to oblige BT to install, keep installed and run the Applicable Systems in order to provide these services. Thus BT is obliged to install and (on OFTEL's interpretation at least) to maintain in good running order all the telecommunication apparatus necessary to provide such services, including not only the apparatus (*e.g.* exchange lines) by means of which messages initially are sent, but all the transmission and switching apparatus (*e.g.* public exchanges) required to ensure their safe delivery.

During the early part of 1987, partly as a result of a strike of its engineers and partly because of the introduction of digital exchanges and attendant changeover problems, particularly in London, BT's quality of service deteriorated; it would appear that the Director General took the view that if he were to take remedial action, this would be based upon BT's obligations under Condition 1 but eventually BT's service gradually began to improve. In the aftermath, at the time of writing, the Director General has issued a consultation paper on BT's standard terms and conditions in relation to contract liability, particularly for fault repairs, and this paper suggests (as the Director General has stated publicly for example in his 1986 Annual Report) that he is considering the case for making BT's performance in this regard an explicit factor in its RPI–3 price control formula.

It is noteworthy that BT's obligation under Condition 1 relates specifically to voice telephony services and telecommunication services consisting in the conveyance of messages; this is important in relation to other licence conditions, particularly Condition 17 (prohibition on undue preference and undue discrimination) which only applies to those telecommunication services provided by BT "in accordance with an obligation imposed by or under" its licence. "Voice telephony" services are not defined, whilst telecommunication services are defined in section 4(3) of the Act as including the conveyance of messages, directory information service and a service consisting in "the installation, maintenance, adjustment, repair, alteration, moving, removal or replacement of apparatus" for connection to a telecommunication system.

International services are dealt with separately in Condition 5.[5]

Condition 46: Simple Resale

In relation to private or "leased line" service, there are two exceptions to BT's obligations to provide service to customers, appearing in Condition 46 (Private Circuits). Under Condition 46.1, in furtherance of the Government's stated policy to prevent simple resale, BT is to take all reasonable steps to prevent the use of its private circuits for the purpose of providing simple resale services. This restriction is to continue until July 1, 1989, when the policy against simple resale is likely to be dropped altogether or modified. Simple resale is discussed more generally at the end of this Chapter.

Under Condition 46.2, BT is excused the obligation to provide private circuits

(a) to another public telecommunications operator whose own licence allows that operator to provide private circuits in the area concerned; and
(b) to an Associate of Hull Council where that Associate intends to provide a "Basic Service" (for example conveyance of data) in BT's licensed area.

Although Condition 46.2 refers specifically only to Hull, it was originally included in BT's Licence particularly to ensure that Mercury Communications would put in its own line capacity and not rely on reselling capacity over BT circuits. It must be debatable whether or not this should have been a matter for concern, given that market forces and competitive pricing could act as a considerable incentive for Mercury to rely on its own facilities. In any event Condition 46.2 does not extend to preclude a Mercury associate from requiring BT to provide a private circuit.

2. Social Obligations: Conditions 2 and 31 to 33 inclusive

Condition 2 (Rural Areas) reiterates the terms of Condition 1 as to universal service, specifically in relation to rural areas. There was sufficient concern expressed in the passage of the Telecommunications Bill regarding rural services and other of BT's social obligations, that pressure was successfully brought to bear to include express mention of such areas.

Special arrangements for the provision of directory information for the blind are referred to in Condition 3 (below). Under Condition 31, BT is required to make arrangements for the provision of telecommunication apparatus which is suitable for the disabled and Condition 32 requires that BT supplies special telephones for those whose hearing is impaired. Under Condition 33, this obligation is extended to require BT to provide apparatus in its public call boxes to assist those people who use hearing aids.

3. Directory Information: Condition 3

Condition 3.1 requires BT to provide directory information services relating to its switched voice telephony services. Such a directory information service is, in the context of "telecommunication service", defined in section 4(3) of the Act as being a service provided *by means of a telecommunication system*, thus leaving outside BT's licence obligations any necessity to supply "hard copy" directories.

BT's directory information services are nevertheless at present provided, free of charge, in both written form (in hard copy directories) and verbally by communication with an operator (who relies on a computerised data base). There is nothing to prevent BT making a charge for such services at any time in the future and clearly the services could also be provided direct to the public by means of an information retrieval system, interactively.

There are special provisions regarding information about the customers of other operators contained in Condition 3.2.

BT must also ensure that its directory information is made available free of charge in a form sufficient to meet the needs of blind and disabled people or, where BT does make a charge for its directory information services and the Director is satisfied that the application of such charges to blind and disabled persons cannot be excluded, BT must pay appropriate reasonable compensation to such persons.

4. Emergency Services: Conditions 6 to 9 inclusive

Condition 6 (Public Emergency Call Services) requires BT to provide to the public free of charge public emergency call services in order to enable people to reach the police, fire, ambulance and coastguard services.

Conditions 7 to 10 require BT to continue to provide other emergency services required by the emergency organisations themselves (Condition 7), to provide maritime services such as the Distress Watch Service (Condition 8), and to make special arrangements in the event of major accidents and other emergencies (Condition 9).

5. International Services: Conditions 5, 47 and 48

Under Condition 5, BT must take all reasonable steps to satisfy reasonable demands for international telecommunication services. This obligation extends not only to the generality of BT's customers but also to other operators, even those competing with BT, such as Mercury. In Mercury's case, access to BT's international routes for Mercury's telephone and telex customers where Mercury had no direct route of its own was secured by the Director General's two determinations.[6]

The meaning of "reasonable demand" and its discussion in the PanAmSat decision is looked at under Condition 1 above.[7]

In order to prevent BT "locking out" competition on international service, Condition 47 prohibits BT from entering into any agreement with an overseas administration which unfairly precludes or restricts similar arrangements by any other PTO.

BT is also prohibited from excluding any other PTO from participating in "international arrangements" into which BT proposes to enter in relation to the installation and operation of any submarine cable linking with any foreign telecommunication system.

Parallel Accounting

Condition 48 (International Accounting Arrangements) requires BT to try to agree with other PTOs, such as Mercury, a Code of Practice on international accounting. In practice, BT and Mercury failed to agree upon such a Code and on March 4, 1987 the Director General duly issued a Code upon the terms determined by him. Accounting rates are the payments one Administration or RPOA makes to another for traffic sent to it by that other Administration. There is also a charge which the Administration sending the traffic renders to its own customers, known for international accounting purposes as the "collection rate."

The Statement issued by the Director General with this particular determination explains government policy on international accounting arrangements. This policy is directed towards ensuring "parallel accounting": this would require BT and Mercury (and any other licensed provider in the future of international telecommunication services) to adopt the same accounting rates, divisions and methods in their dealings with foreign carriers in order to avoid "whipsawing," the practice whereby a monopoly foreign carrier would be able to try to negotiate cut rates with one United Kingdom operator to the detriment of another.

Parallel accounting might be viewed in principle as an anti-competitive arrangement, but United Kingdom finds favour with the regulatory authorities because it protects competition between United Kingdom international operators and is also justified in its protection of the national interest or, as the Director General puts it, "the United Kingdom bargaining position."

In some instances competition will be better served by allowing exceptions to the general rule of parallelism, particularly where whipsawing is unlikely, and the determination therefore permits this in certain cases, for example where there is more than one international operator in the foreign country concerned (this would include the United States and Japan). This is subject to the Director General being satisfied that the interests of the relevant operators and of users in the United Kingdom would not be prejudiced by such an arrangement.

An exception to parallel accounting is also made, subject again to the Director General being satisfied, for the situation where an operator has been unable to obtain arrangements for proportionate return with an overseas operator. Proportionate return is intended to secure for the sender of traffic the return to him of a proportionate share of incoming traffic to the United Kingdom. The economics of international service are such that it is vital for the overall profitability of an operator's services to obtain the revenues arising from a share of the accounting rate (sometimes called the accounting rate "in-payment") arising from such return traffic.

Under Condition 48.2, where the Director General is of the opinion that BT is about to enter into or vary an agreement or arrangement with a foreign administration which establishes international accounting methods, rates and divisions in a way which would prejudice the interests of other providers of international telecommunication services, he may issue a direction prohibiting BT from going ahead with the proposal.

6. Maintenance and Fault Repair: Conditions 4, 10 and 44

Maintenance. In support of its universal service obligation, BT must also in certain circumstances provide "Maintenance Services." These services are closely defined in Schedule 1, Part 1 and include not only carrying out repairs but also "any activity involving the removal of the outer cover of the apparatus." Excluded are operations incidental to the installation, "Bringing into Service" (again closely defined) or routine use of the apparatus.

BT's obligation to provide Maintenance Services for its customers does not apply in a variety of situations, for example where the system or apparatus to be maintained is beyond repair, or necessary components are not available or where it has not been supplied by BT and it is not a term of the licence for such system or apparatus that BT should provide such services, or where its approval under Section 22 of the Act for connection to BT's systems does not require it to be maintained by BT.

Fault Repair. Any PTO is free to provide such levels of fault repair service as it finds commercially expedient. However, in BT's Licence (Condition 10) emergency services are understandably singled out for special treatment in this respect. Any person or categories of persons or authority:

(a) engaged in the provision of an emergency service to the public, or the provision or supply of essential services or goods, or in public administration;
(b) whose name is notified by the Director General to BT and who pays BT's charges for the service; and
(c) who has a bone fide need for an urgent repair,

is entitled to such repair by BT in order that disrupted service may be restored "as swiftly as practicable."

This priority fault repair service is to be available for 24 hours a day or such lesser a period as BT may agree with the relevant person. At present BT's charges for emergency service priority fault repair are £16.50 per exchange line for round the clock service and £6.00 per annum for service only during working hours. BT also provides to all its customers, by contract, its "premium maintenance services" at varying rates for different response times.

In any case where another party's licence for a telecommunication system or apparatus requires that BT should provide maintenance services (for example in relation to NTTA), Condition 44 obliges BT to publish its charges and other terms and conditions for such maintenance services and to supply such services at those charges and upon the stated terms and conditions, with very limited exception.

7. Charges, Terms and Conditions: Condition 16 and 24 to 26 inclusive

Condition 16 requires BT to publish its charges (or a method for determining its charges) and other terms and conditions on which it offers to provide its various telecommunication services and not to depart from such charges, terms and conditions. It should be noted that this requirement only extends to telecommunication *services* and ancillary activities which BT is *obliged* by its licence to provide and undertake. Thus the supply of telecommunication apparatus is not regulated in this way. Moreover the condition does envisage the possibility of the Director General excusing BT from compliance with its provisions in certain respects.[8]

In the author's view, the obligation not to depart from published terms and conditions should not prevent the application by BT in a particular case of additional terms or conditions which do not conflict with or materially affect the published version, provided these additional provisions are warranted by the special circumstances of the case or of the customer and the customer is not thereby unduly preferred or discriminated against.

Publication is to be effected by sending a copy of the charges, terms and conditions to the Director General, at least 28 days or, in the case of VADS-type services ("Supplemental Services Business") at least one day, before any amendment is made, and by placing a copy in the office of the General Manager of each of its telephone areas and by sending a copy to any person who so requests. By contrast, in relation to any type of service, Mercury need only give a minimum of one day's notice to the Director General of any amendment to its charges, terms and conditions.

These filing requirements are purely administrative; PTO charges *per se* are not subject to approval by the Director General, or any other form of direct control[9] except, in BT's case, with respect to the services covered by the RPI–3 formula.

RPI–3 Formula

This formula (initially to continue in force for five years) in Condition 24 is a ceiling on BT's price increases for inland services referenced by a basket of prices for business and residential rentals, local calls and trunk calls. Price increases for such services overall must be no more than three percentage points less than the annual rate of inflation. If price increases in any year are less than this limit, BT can take the credit in either of the next two years.

To date, BT's various price changes have come into effect in November of each year. Such changes often provoke questions from the public as to whether BT is actually complying with its licence obligations and the RPI–3 formula and whether in fact that formula is the correct method of price control. Accordingly the Director General has made studies of these price changes and his last report, published in November 1986, analysed BT's tariff changes which came into effect that month. This analysis shows that although BT's prices for certain services, such as exchange line rentals, local calls and short distance national calls had risen, there had been reductions in other services, such as long distance national calls; thus the overall net effect was that taking into account the under-utilisation of permitted price increases in previous years BT had made a small average price reduction and had met the RPI–3 requirements.

Although not required to do so by its licence, the Board of BT has given an undertaking, which was referred to in BT's 1984 flotation prospectus, to limit increases in the price of domestic exchange line rentals to two percentage points above the rate of inflation (RPI–2).

At the end of the five year period for which the RPI–3 formula applies there will be no further restriction on

BT's prices, in the absence of a licence modification for its continuance or replacement by an alternative. In commenting on this aspect during the passage of the Telecommunications Bill, the then Minister for Information and Technology stated that if by that time competition had not been extended significantly to the area of local networks the Director General might decide to extend the price control. Without BT's agreement he would have to refer the necessary licence modification to the MMC for it to judge whether continuation of the control was or was not in the public interest. In January 1988 the Director General issued a Consultative Document inviting comment on continuation of some kind of formula for controlling BT's prices.

Condition 25 requires British Telecom to levy uniform maintenance charges for its exchange lines throughout the country, over the first five years of the licence. Similarly, Condition 26 provides that residential connection charges will be uniform over the same period when the work involved takes less than 100 hours.

8. Competition—Fair Trading: Conditions 17 to 22 inclusive, 35, 36 and 39

The fair trading conditions in BT's licence are selective in that they are aimed at particular anti-competitive practices, which may not be exhaustive. It is the author's view that the competition provisions of the regulatory regime, primarily embodied in PTO licences such as BT's, would have been considerably strengthened by a "catch-all" licence condition along the lines of a domestic equivalent of Articles 85/86 of the Treaty of Rome.

For example, predatory pricing which was anti-competitive in its effect is not expressly covered in any of the fair trading conditions. It is quite possible that Condition 17 would not be applicable to such pricing in all instances for that condition is geared to a preference to or discrimination against BT's customers rather than addressing the effects on its competitors.

Condition 17 provides that BT shall not show undue preference to, or exercise undue discrimination against, any persons with respect to its various services. Significantly, in the case of its telecommunication services the condition applies only to those services which BT is by its licence obliged to provide.

Any question as to whether or not BT has indulged in undue preference or undue discrimination is to be determined by the Director General—Condition 17.3.

There is no prohibition on discrimination or preference as such—the test is whether it is "undue." The Oxford English Dictionary meaning for "undue" includes "improper," "unreasonable" and, in particular "excessive." Further, in the House of Lords decision in the *South of Scotland Electricity Board* v. *The British Oxygen Company*[9] Lord Keith stated that the word "undue" encompassed not only illegitimate reasons but could also mean "excessive."

He also considered that it was a matter of fact and degree, in effect a jury question.

In their further appeal in the same action[10] their Lordships went further into the scope of the phrases "undue discrimination" and "undue preference" and, for example, dismissed the Electricity Board's argument that the complainant would have to show there was another customer paying less in order to begin to prove undue discrimination. The decision is also interesting for its discussion of the relevance of costs to discrimination, Lord Merriman stating, for example, that the phrase "shall not exercise any undue discrimination" had to be considered as a whole and that "a fair distribution of the cost of supplying electricity as between one class of users and another is not to be left out of consideration any more . . . 'than any other circumstance which would affect mens' minds' ". This case is also useful for its review of many of the older cases including particularly those under various Railway Acts, where, again, the prohibition on "undue preference" appeared.

The British Oxygen cases mentioned above concerned section 37(8) of the Electricity Act 1947, which prohibits area electricity boards, in fixing tariffs, from showing undue preference or exercising undue discrimination against customers. Over the years Electricity Board cases have therefore produced a number of useful decisions on the meaning of these phrases.

Given the litigation which has been seen with regard to Electricity Boards and, indeed, the Water Authorities, whose governing legislation[11] also contains prohibition on undue discrimination and undue preference, it is a little surprising that there appears to have been little challenge of BT's charges under Condition 17, although the answer may be that many of these have been disposed of by the Director General determining in the first instance that BT had not breached its licence. The first example of which the author is aware and where the Director General appears to have found against BT is the William Hill case,[12] where the Director General determined that BT had provided the William Hill Raceline Service at a lower rate than that available to other Premium Service Providers.

The presence of undue discrimination or preference should thus fall to be determined by the Director General in accordance with the facts and by applying relevant legal principles which have emerged from the cases. If and where the Director General establishes that such discrimination or preference is present he should consider whether or not to take action in accordance with his duties in section 3 of the Act, particularly 3(2) which deals specifically with competition.

Undue preference and undue discrimination can be established if BT unfairly favours to a material extent any part of its own business, where this places its competitors at a significant competitive disadvantage—Condition 17.2.

Condition 17 does not apply to a value added or data service able to be run legitimately under the Branch Systems General Licence (see Condition 17.4(b)).

Condition 18 (Prohibition on Cross-Subsidies) provides that the Director General may step in to prevent or give other directions to BT where it is unfairly cross-subsidising its apparatus supply business, its production of telecommunication apparatus, its provision of mobile radio services or its "Supplemental Services Business." Any material transfer between any part of its businesses and these other businesses must be recorded at full cost in its accounting records ('full cost" being defined as including the market rate of interest for the money transferred). "Unfairly cross-subsidising" is, however, not defined and is left to the Director General to interpret. Subsidisation within BT's Systems Business or from another of its businesses to support the System Business is not prohibited.

Condition 19 (Access Charges) is dealt with below in section 9.

Condition 20 (Separate Accounts, etc.) obliges BT to maintain separate accounts in relation to its Systems Business, its apparatus supply business and its Supplemental Services Business so that these can all be assessed and reported on separately. BT's accounting statements must set out the revenue and financial position of each of those businesses and include details of items charged between the various BT businesses or apportioned between them.

Condition 21 (Apparatus Production) provides that if at anytime BT becomes engaged in the business of production of telecommunication apparatus (to date BT has mainly bought in apparatus from manufacturers such as GEC and Plessey) it shall carry on such business through a subsidiary. BT has in fact transferred its own apparatus production business to one or more subsidiaries. In this connection, in 1986 BT acquired the apparatus production company, Mitel. This acquisition was the subject of a report[13] by the MMC after a reference to the MMC by the Secretary of State under sections 69(2) and 75 of the FTA. The report itself is of interest for its discussion of the United Kingdom telecommunications market and its development thitherto.

The MMC found that the proposed merger could be expected to operate against the public interest and by a majority recommended that it should not be allowed unless BT gave certain undertakings on its future conduct in relation to Mitel, which it suggested should be monitored by the Director General and possibly incorporated in BT's licence. Subsequently the Secretary of State allowed the acquisition to proceed subject to BT giving certain undertakings. These included restrictions and controls on BT's acquisition of Mitel equipment.

Condition 21.4 requires that, if so required by the Director General and where BT is a monopoly purchaser, BT should adopt an open tendering procedure before acquiring any apparatus from its own apparatus production company. The remaining provisions of this condition describe the open tendering procedure and specify various exceptions to this rule (see 21.6) as well as giving guidance to the Director General as to the matters to which he should have regard in exercising his discretion under this condition.

Condition 22 (Prohibition of Preferential Treatment) requires BT to give other suppliers of apparatus similar facilities in any area where a person engaged in BT's systems business delivers or connects apparatus as an incidental of carrying on that business; similar equality of treatment must be afforded where a person normally engaged in the apparatus supply business of BT arranges for BT's systems business to install apparatus or provide certain services. Clearly, in its ability to combine the supply of apparatus and services BT is potentially at a competitive advantage and this Condition is an attempt to try to restore some balance for BT's competitors. However it does not necessarily afford protection in the converse situation, namely to providers of telecommunication services competing with BT and who require the supply, for example, of additional apparatus by BT in order to complete their service for a customer.

Condition 23 (Alterations to the Applicable Systems) is designed to protect both BT's suppliers and its competitors in that if BT is about to change its systems or apparatus in a way which would necessitate modifications to systems or apparatus or even lead to equipment redundancy, BT must first notify the Director General and also follow a procedure for prior consultation and advance warning to suppliers affected. Such a procedure has been settled.

Condition 35 (Prohibition of Linked Sales) prohibits what are normally termed in competition law parlance "tie-ins." Specifically BT is prohibited from requiring a customer for any particular service or apparatus to take any other service or apparatus from BT as well, unless there is a good technical reason for this.

The same condition contains a restriction on "packaging" services for the same purpose, where the terms or conditions are more favourable than would apply if the additional service or apparatus were not provided, unless the Director General agrees otherwise. The prohibition is relaxed in Condition 35.3 to enable BT to offer quantity discounts, for example.

Condition 36 (Prohibition of Certain Exclusive Dealing Arrangements) is a strangely conceived and worded condition, preventing BT, without the written consent of the Director General, from imposing extraneous restrictions on its suppliers, with respect both to supply of unrelated apparatus or services and to intellectual property rights.

The aim of this condition is perhaps more accurately concentrated in Condition 36.2, where a requirement of the grant to BT of sole rights in respect of the supply of customer apparatus can be prohibited by the Director General where he is satisfied that the suppliers concerned are not genuinely willing to confer such rights on BT.

Unfortunately, Condition 36.4 contains so many exceptions to these rules that much of their good is undone. For example, BT is free to require that "other" telecommunication apparatus be supplied or some other telecommunication service be provided by the supplier where that other apparatus or service "is reasonably related to that supply or provision": the terms in which this has been drafted are too wide and it would have been preferable had the line normally taken by, for example, the European Commission been followed; reference could then have been made to items which are technically indispensable to the proper exploitation or use of the apparatus or service.

Again under Condition 36.4(c) BT is permitted to

require the transfer to it of intellectual property rights which the Director General agrees "is necessary or desirable to facilitate the running of any of [BT's] Systems": admittedly, the Director General's discretion is retained and may give sufficient protection against abuse but the apparent authority given to any attempted takeover of a patent or copyright holder's rights seems, on the face, unnecessary and even unjustifiable: possibly the draughtsman meant only to refer to the *grant* (*i.e.* licence) of intellectual property rights rather than their complete disposal, but the wording is ambiguous. A further paragraph (f) in Condition 36.4 states that BT is free to require the transfer to it of any interest in intellectual property "to the extent that that is reasonably necessary for the purpose of enabling [BT] to secure alternative sources of supply of telecommunication apparatus": again, a term which, although of no direct legal effect on anyone other than BT, might encourage the perception that BT had a right to demand such a transfer. Indeed generally perhaps the best that can be said for these intellectual property provisions is that of themselves they are not enforceable against the legal owners of the rights, who are not bound in any way by the terms of BT's Licence.

Condition 39 (Intellectual Property) provides that the Director General may give directions to prevent BT using intellectual property rights to restrict the availability of any product where this is liable to prevent the connection of any telecommunication system or apparatus or the provision of any service. The Director General is also empowered to direct BT to grant intellectual property rights to other persons in order that such connections or services may be made or provided. There are saving provisions where any requirement on BT would result in it breaching the terms of any licence or assignment of intellectual property rights. These provisions can be contrasted with other industries under monopoly control where compulsory licensing procedures are not generally available. This Condition may therefore be used to prevent BT from withholding the supply of apparatus, but only where this is contrived by the use of intellectual property rights. A more pro-competitive licence condition might have extended this principle to any refusal to supply whether or not the product is, say, patented or then subject of copyright. However, the Competition Act may be operable in such circumstances.[14]

9. Interconnection—Connection of Other Systems: Conditions 13 to 15 inclusive

Condition 13 (Connection of Systems providing connection services) is one of the most important conditions in BT's Licence affecting liberalisation. It secures for other PTO's, the right to require an interconnection agreement with BT in order that messages may pass from the other operator's system, into BT's system, typically to reach BT customers and in order also, where the other PTO is a long-line operator (*i.e.* Mercury for present purposes), that customers of both BT and that operator may exercise freedom of choice as to the system by which their messages are conveyed. Essentially, these principles, to be found in Condition 13.1(a) and 13.1(b) respectively, represent two of the main objectives of liberalisation, "any to any" (the ability of any customer of one public operator to call the customer of any other public operator) and "customer choice."

The preamble to Condition 13.1 states that unless it is impracticable to do so, BT must enter into an agreement with an operator for these purposes if that operator is licensed to run a "Relevant Connectable System" and requires BT to do so.

A Relevant Connectable System refers to a system which is licensed to be connected with BT's and to provide services for reward to the public but which is *not* run under a class licence for the connection of which BT offers standard terms and conditions satisfying the requirements of Condition 16, although the Director General has the discretion to exclude a system which would otherwise be a Relevant Connectable System for these purposes.

The procedure under Condition 13 is that the operator

must seek agreement with BT on a number of matters specified in Condition 13.4 and if "after a period which appears to the Director to be reasonable" (in the particular cases which have arisen so far this has been of the order of six months) an agreement has not been achieved, either party may apply to the Director General for him to determine the terms and conditions of their proposed agreement. The criteria which the Director General must bear in mind for this purpose are set out in Condition 13.5 and, in relation specifically to longline PTO's (for example Mercury), in Condition 13.6.

Condition 13 is silent upon how soon after the Director General's determination an agreement should have been entered into, incorporating the terms and conditions so determined but it can be assumed that the parties thereafter have a reasonable period of a few months in which to do so and if BT were to default in this (the other operator being willing to sign) it would be in breach of its licence.

Enforcement of the Interconnection Agreement itself is underpinned by Condition 13.8 which allows the Director General to step in where BT is failing to carry out connection work and issue a direction requiring BT to do whatever is necessary to remedy the default. However, 13.8 is quite inadequate to provide recourse to the Director General to ensure due performance of all aspects of an Interconnection Agreement. It should not be forgotten, however, that an Interconnection Agreement constitutes a contract like any other and is therefore enforceable through the courts in the usual way.

The first occasion on which the Director General was called upon to exercise his powers under Condition 13 occurred in the dispute between Mercury and BT over the terms of their Interconnection Agreement for telephony, which was eventually referred to the Director General in December 1984. The determination proceedings were briefly delayed awaiting the outcome of litigation between BT and Mercury over heads of agreement which had been signed by the parties the previous summer, before the enactment of the Telecommunications Bill and before the issue of BT's Licence. In this litigation, which finally came before Mr Justice Leggatt in the Commercial Court in March 1985, BT had claimed a declaration that in any agreement made between the parties pursuant to Conditions 13.1. and 13.4 of BT's Licence, BT was entitled to require that all the terms and conditions of the heads of agreement should be incorporated in such new agreement.

BT's claim was struck out, the judge finding that the heads of agreement were not legally binding and therefore did not prevent Mercury's application to the Director General from proceeding in relation to all the matters which Mercury claimed were in dispute.

After this brief hiatus, the Director General proceeded with his determination which was eventually issued on October 11, 1985. The text of the determination, which is a public document, laid down binding terms and conditions, which eventually in the early part of 1986 were incorporated in a formal agreement between the parties, on the connections to be made between BT's telephone system and Mercury's system; it included the time limits for such connections, the charges for connections, call charges, international service and a number of technical matters including quality of calls and numbering arrangements.

The Director General commented on this first determination in his 1985 Annual Report, as follows:

> "One of the main issues I had to determine concerned the flexibility that customers should have in choice of route; for example, should BT customers be allowed to choose to route calls over the MCL [Mercury] network regardless of the distance of the origin of the call and the ultimate destination from nearest points on the MCL network, or should some distance-related limit be imposed? And should MCL customers be allowed to use BT's system for making international calls?
>
> I decided that BT should have the obligation to provide connections to the MCL system at local

> exchanges and at trunk exchanges for use without limit, except such limit as was required to ensure that messages had a satisfactory quality; I also decided that MCL should be permitted to transfer calls to BT for connections to other countries. I established prices that were consistent with the licence rules and in particular that gave MCL reasonable incentive to extend its system, rather than relying unduly upon that of BT."

This determination is particularly noteworthy in respect of the regime established for call charges between competing operators. In the pre-determination negotiations BT had been pressing for call charges geared to its standard customer tariff, whereas Mercury had pressed equally strongly for a cost-based charging system on the premise that under Condition 13.5 this would be one of the principles to be secured by the Director General in making his determination. The eventual determination did, indeed, take this line in relation to inland calls, although applying a differential charge depending on the extent of use of BT's systems; international calls were treated quite differently, with Mercury being left to pay BT's standard customer tariff. The connection charges (for connection of Mercury's system to points established at BT's trunk exchanges) which Mercury is liable to pay are also quite different from BT's normal customer arrangements.

The determination established an agreement between BT and Mercury for some considerable longevity but in reproducing the determination the agreement contains provision for review by the Director General at any time after two years from the date of its signature. The earliest this review could take place was therefore March 1988. Any such review would be dependent upon the Director General being convinced that there has been a material change in circumstances or that the intended competitive effects of the Agreement were not being met; he could then make a further determination to amend the agreement for this purpose, again on the basis of the criteria set out in Condition 13.

The second and so far only other determination by the Director General under Condition 13 was in relation to the terms and conditions to apply in an agreement between BT and Mercury concerning international telex calls.[15]

Condition 14 (Connection of Other Systems and Apparatus), by contrast, does not apply to Relevant Connectable Systems but imposes upon BT an obligation to connect all other parties' systems to its own system where these systems are licensed to be so connected. BT is similarly obliged to connect apparatus to its system where such apparatus is duly approved for this purpose. There are, as usual, exceptions to BT's obligations, in particular where the apparatus did at one time, but does no longer comply with the relevant standard or standards. Also where, although conforming to this standard the apparatus is liable—in BT's opinion—to cause death or personal injury, or damage or materially impair the quality of telecommunications service provided by BT, and the Director General has not expressed a contrary opinion.

This principle of "material impairment" promises to be quite important in the apparatus approval regime as well as in relation to connection of private networks and systems to those of PTO's. Current practice of OFTEL, rightly or wrongly, appears to be that in relation to voice telephony the minimum quality standards set out in the provisional NCOP (Network Code of Practice)[16] are to be taken as the threshhold for material impairment.

Condition 15 permits a person, who is authorised to provide telecommunication services to others, to provide such services whilst its system is lawfully connected to BT's systems.

Access Charges—Condition 19

Condition 13.10 states that it is without prejudice to this Condition, although due account was to be taken for the purposes of Condition 13 of any access charge imposed on the operator requiring interconnection.

Condition 19 specifically empowers BT to impose access charges, the proceeds of which are to be used exclusively to defray certain costs specified in Condition 19(c)(ii). To date, no such charge has been imposed, but the Director General's November 1986 report on BT's tariff charges indicates that he is beginning to consider the matter. For example, in that report he points out that an access charge could be used to defray BT's losses on exchange line rentals.

10. Numbering: Condition 34

Every branch system or terminal apparatus has a number or numbers which identify the relevant Network Termination Point and which must be incorporated in a numbering plan of the PTO having direct connection with that system or apparatus.

PTOs must also adopt and allocate numbers for identifying other PTOs and other operators of systems and providers of services as well as providing numbers for access to other operators, for example to facilitate choice.

Under Condition 34 BT is obliged to adopt a numbering plan which describes the method adopted for allocating numbers in respect of its Network Termination Points in order that messages may be conveyed to apparatus or systems connected to those Network Termination Points (Condition 34.2).

BT is obliged to consult with the Director General regarding its numbering plan and this consultation procedure is now underway. More significantly BT is obliged under Condition 34.5 to provide the Director General with proposals for developing the numbering plan in order to secure certain essential features of liberalisation such as: sufficient numbers being available to meet demand; numbers including as few digits as practicable and their allocation not conferring any undue advantage on BT or disadvantage to its competitors; the cost of changing BT's systems to accommodate the plan being reasonable and any inconvenience caused by any alteration of the numbering plan being minimised (Condition 34.5).

The numbering plan adopted by BT must also be compatible with the numbering arrangements applied by other PTOs and the Director General may issue his determination to this end, taking account of the criteria set out in Condition 34.8.

Thus numbering arrangements, before 1984 under the effective responsibility and control of BT, are now in principle controlled by OFTEL although until OFTEL have, in consultation with interested parties, resolved the compatibility of the various operators' numbering plans there is obviously a danger that numbers may be squandered, misused or used in such a way as to exhaust their availability as, as many would describe it, a national resource. Under Condition 34.10, although it is stated that BT should not charge a person for allocation of a number (unless it is a "coveted" number allocated to a non-PTO on request), BT is nevertheless not precluded from recovering the reasonable cost of allocating a number and the reasonable cost of carrying out changes to its systems to ensure that messages reach that number.

BT is obliged (under Condition 34A.1) to adopt a separate numbering plan for its Supplemental Services Business.

11. Testing: Conditions 40 and 41

Testing is a statutory requirement for the purpose of approval of apparatus under section 22 of the Act (on apparatus approval generally see below).[17] Condition 40 applies to non-statutory testing requirements, in effect tests required or carried out by unauthorised persons. In such circumstances, where BT is imposing such requirements, the Director General can require that BT should desist.

Under Condition 41 where BT (which would in practice mean its "Teleprove" unit) carries out any test or assessment of telecommunication apparatus it is to keep confidential any information obtained in the course of such work and may not disclose it to anyone, including its own employees. The main exceptions to this are where consent is obtained from the Director General, from the producer or supplier of the apparatus or the person who requested BT to carry out the test or assessment.

12. Wiring: Conditions 42 and 43

Use of wiring is important for liberalisation and the penetration of competition. Before 1981 BT's systems embraced wiring and apparatus on customer premises and there was no concept of a network boundary at the interface of the public system with the branch system, as there is today. An historical overhang from this arrangement is that BT owns much of the wiring in customer premises throughout the country. A customer may wish to make use of this wiring to enhance or reconfigure his system; if it cannot be worked on independently of BT's system, delays and complications could result.

Accordingly Condition 42 restricts BT's installation of integrated wiring in the future and secures unencumbered access to that wiring on reasonable terms where it already exists.

Condition 42 provides that BT shall not, except in limited circumstances, install a line on customer premises in such a way as to prevent access to the wires or cabling of any other system on such premises. In relation to existing premises BT is also obliged to install additional apparatus in order that a system on the customer's premises may be run by someone other than BT and so that operations may be carried out on other systems separately from BT's systems.

Under Condition 43 BT is obliged to make available (*i.e.* typically, rent) to any person, any of its non-system apparatus (for example, wiring) for that person to use in the running of a telecommunication system, typically a branch system. This is to be done on terms and conditions and subject to charges which are no less favourable to the user than would apply if BT ran the system itself, provided the maintenance services or supplied telecommunications apparatus comprised in the system. In a case where BT does not retain ownership or control of the apparatus (*i.e.* sells it to the applicant) it is to be made available "at a reasonable capital charge." The practicability of performance of this obligation is uncertain.

Under Condition 43 BT must permit the user of telecommunication apparatus to carry out work on such apparatus except in the case where that apparatus is comprised in BT's systems in such a manner that Maintenance Services (referred to in Condition 4) cannot be carried out independently of BT's own operations on its systems. The apparatus to which Condition 43 applies is fully described in Condition 43.3.

Up until 1986 OFTEL had received numerous complaints about the restrictive effect on competition in apparatus supply resulting from BT's ownership of most internal wiring systems. In particular, notwithstanding Condition 43, in practice BT had followed a policy of offering wiring only for sale. Consequently a user who wished to buy call routing apparatus from one of BT's competitors would first have to pay for such wiring whereas a user who bought new apparatus from BT could continue to rent the wiring. This was acting as a disincentive to purchase from BT's competitors and the matter was investigated by the Director General with a result that BT agreed to new arrangements to try to overcome these problems. Under these arrangements:

(a) BT agreed to offer Maintenance Services for call routing apparatus supplied by its competitors (where such apparatus was of a type already being maintained by BT in the district concerned) provided doing so was commercially viable. In such cases where the customer entered into a contract for maintenance with BT, he would be able to choose whether to rent or buy the wiring. (Note here that the BSGL requires that wiring and call routing apparatus be maintained by the same entity; accordingly where BT owns the wiring it would follow that it must also maintain the call routing apparatus.)

(b) If any upgrading of wiring is necessary, users would be charged on an equal basis whether they own or rent the wiring.
(c) Where BT has provided separate wiring in order that customers may obtain call routing apparatus and/or Maintenance Services from BT's competitors, no immediate lump sum charge would be made. Instead there would be standard connection charges as new exchange lines are provided, as well as standard removal or re-termination charges for existing circuits which are reprovided through that separate wiring.

Overall the effect of these new arrangements was intended to balance out rental charges and purchase prices for wiring so that a customer considering taking new apparatus would not be influenced in his decision by whether or not he should buy or rent.

13. OSI Access to Supplemental Services Business: Condition 40A

Essentially this requires BT to provide means of access to Relevant Services, falling within its Supplemental Services Business, conforming to OSI standards, within 12 months of the date on which an OSI standard is specified by the Director General. This requirement mirrors the same provision in the VADS Class Licence.

So far as the author is aware, no OSI standard has yet been specified by the Director General. There are also certain limitations to BT's obligations as set out in Condition 53.[18]

14. Customer Confidentiality: Condition 38

Under this condition BT is obliged to operate a Code of Practice restricting the disclosure of information about its customers by its employees engaged in various BT businesses, each of which has a separate code. The code for BT's Supplemental Services Business is in similar terms to the model code issued by the Director General under the VADS Class Licence.

15. Disputes etc; Code of Practice: Condition 27

Under this condition BT is also obliged to issue a Code of Practice giving guidance to BT's customers and employees regarding disputes and complaints. The Code was first issued in November 1984 and contains sections on service, phone books, bills, operator services, fault repairs, payphones, complaints and arbitration.

Arbitration

As the Consumer Code of Practice makes clear, BT is obliged to include in its standard terms and conditions an inexpensive independent arbitration procedure for resolving disputes which do not involve a complicated issue of law or a sum greater than a particular sum specified by the Director General from time to time (presently £1,000). The arbitration procedure adopted by BT (and Mercury) is set up under the auspices of the Chartered Institute of Arbitrators.

16. Joint Ventures: Condition 49

BT is obliged to notify the Director General at least 30 days before its entry into particular agreements or arrangements with third parties as follows:

(a) for the running of a telecommunication system requiring a licence; for providing telecommunication services or for the production of telecommunication apparatus where that production would lead to a monopoly situation;
(b) for the establishment of a partnership for such purposes and in such circumstances; or
(c) for a joint venture to run a telecommunication system requiring a licence or to provide such telecommunications services.

Any such agreement for establishing or controlling a body corporate as referred to in (a) above or for establishing a partnership as in (b) above only applies where BT has or is to have not less that 20 per cent. of the voting power in the particular entity.

It was not because of this condition that the BT/IBM abortive JOVE venture for value added and data services came to the Government's attention, but rather because of its licence application.

17. Associates: Condition 50

This Condition prevents BT avoiding its obligations by using another member of its group ("Associate") in that the Director General can intervene to issue directions to BT requiring it to take the necessary steps to ensure that its Associate ceases or otherwise remedies the default.

18. Value Added and Data Services

In providing any service as part of its Supplemental Services Business (equivalent to a value added or data service which can be provided by a non-PTO under the VADS Class Licence) BT is obliged to abide by the same conditions as apply to major service providers under the VADS Class Licence.

19. Exceptions and Limitations to BT's Obligations: Condition 53

As has been seen, BT's obligations are not absolute. This particular condition excuses BT from any obligation to provide service in the event of *force majeure* circumstances (Condition 53.3) and it is also not obliged "to do anything which is not practicable" (Condition 53.2).

There are further relieving provisions applicable only to voice telephony service. Here Condition 53.4 provides that in specific circumstances BT shall not be under any obligation to provide such service, for example where there is no reasonable demand or where the necessary apparatus is not available. For non-voice services Condition 53.5 provides in addition that BT's obligation may not apply where provision of service is not economic.

Another potentially significant exception is in 53.6, which excuses BT from being obliged to provide service where the customer is in breach of his contract with BT, or refuses to enter into a contract with BT (unless BT is behaving unreasonably in this respect), or is using apparatus illegally or has dishonestly obtained service from BT.

In the case of BT's Supplemental Services and in particular the Condition 40A obligation to provide means of access conforming to OSI standards, BT is excused where it is unable to comply because of non-availability of anything necessary for such purpose or, because BT cannot, through no fault of its own, install necessary apparatus. The Director General may also dispense with BT's compliance, where interests of consumers would not be promoted to any material degree by the introduction of OSI standard.

The remainder of this Condition preserves for BT a number of rights and discretions which an operator would wish to be able to include in its conditions of service.

Simple Resale

We have already seen under Condition 1 above that in relation to private service Condition 46.1 requires BT to take steps to prevent the use of its private circuits for the purpose of providing simple resale services. Simple resale is defined in Condition 46.3 as being a service whereby messages are conveyed from a public switched network over a private circuit and then into the same or another public switched network (an exception being made for elements of this routing forming part of a Centrex System—see Chapter 7[19]). This would have been an obvious way for operators to provide a cheap service alternative to that of the PTOs, simply by leasing raw capacity from the PTOs and reselling service using all or part of that capacity.

The VADS Class Licence now authorises resale for the purposes of data services, but this right does not extend to simple resale as such.

Traditionally tariffs for private circuits have been below cost and therefore were it not for the simple resale restriction, any number of additional networks might have built up to compete with the PTOs using private circuits linked together as the operator's network. Gradually private circuit tariffs are being rebalanced and it may be expected that by the time the restriction is to end in July 1989, tariff levels would have been adjusted sufficiently to ensure that resellers cannot unfairly exploit this market to the detriment of PTOs. It may even transpire that differential private circuit tariffs are applied by PTOs, one level for ordinary customers and the other level for resellers, perhaps by reference to traffic volumes. Clearly the Director General should require to be satisfied as to the fair computation of these charges and in particular assure himself that they do not amount to undue discrimination contrary to Condition 17.

Notes

[1] At page 37 above—Enforcement of Licence Conditions.
[2] Now CBA s.4.
[3] At p. 77 below.
[4] Statement of Director General, March 18, 1988.
[5] At p. 53 below.
[6] See p. 66 below.
[7] See p. 49 above.
[8] See opening words of Condition 16.1, B.T. Licence.
[9] [1956] 1 W.L.R. 1069.
[10] [1959] 2 All E.R. 225.
[11] Water Act 1973 as amended by Water Charges Act 1976.
[12] *The Times*, Saturday March 19, 1987.
[13] January (1986) Cmnd. 9715.
[14] See Chap. 9 below at p. 143.
[15] July 31, 1987.
[16] See Chap. 8 below at p. 131.
[17] See Chap. 8 below at p. 125.
[18] At p. 77 below.
[19] At p. 112 below.

Chapter 5

BRANCH SYSTEMS GENERAL LICENCE

A. Scope

1. This is the Class Licence applicable to the vast majority of fixed telecommunication systems in that it covers everything from telephone handsets and internal wiring (whether at home or in the office) up to the junction with the PTO network termination point, to very large systems incorporating PBX, peripherals and their associated wiring again up to the point of connection to a PTO's NTP. The BSGL does not apply to the running of mobile radio apparatus other than a cordless telephone working with a handset comprised within the branch system.

OFTEL has issued a useful booklet on the BSGL entitled "Explanatory Guide to the Class licence for the Running of Branch Telecommunication Systems." No application need be made, nor is it necessary to register or pay any fee, in order to enjoy the benefit of the BSGL.

2. The first 11 "General Conditions" are common to the BSGL, the VADS Licence and individual licences based on these class licences but are most often encountered in relation to the BSGL and for that reason will be dealt with here.[1] As for the remaining nine conditions and the BSGL Schedules, for ease of reference and in parallel with the analysis of other licences elsewhere in this book, the most important substantive provisions of this licence are dealt with in discrete sections, looking in turn at systems licensed under the BSGL, services which can be provided by branch systems and the conditions applying to connections to branch systems, such as connection of private circuits, including the General Conditions.

3. In relation to the BSGL but also to all other telecommunications licences there is an important general point as to licence scope and the licence appropriate to a par-

ticular system. This is that every licence granted under the Act has its own integrity: by this I mean that the licensee is required to comply with all its terms and conditions if it is to apply to him and cannot rely in part on one licence and in part on another in order to validate the running of a particular system. Only one licence (class or individual) can therefore be applied to a particular system at any one time.

B. SYSTEMS LICENSED[1]

1. What is a "branch system" as covered by the BSGL? Rather than tell us what it is specifically, the Licence adopts the easier drafting approach of stipulating that it is any system of whatever size and purpose, however it is run, and then effectively eliminates from this catch-all various types of system according to their location and the services provided. As explained below, these licensed services are described in Schedule 3.

2. The preamble of the BSGL contains the formal grant of the licence whereby permission is given to all persons (unless the Secretary of State or Director General has revoked the licence in relation to any of them) to run a telecommunication system as described in Annex A. The effect is to authorise the running of any kind of telecommunication system, including one conveying speech and data messages, visual images, including facsimile and video conferencing, telemetry and telecontrol signals.

3. The system must be located within England, Scotland, Wales and Northern Ireland, excluding the Channel Islands and the Isle of Man (this territory could eventually be extended to offshore structures on the continental shelf but beyond the United Kingdom 12 mile limit). The licensed system must be run only in premises occupied by the licensee, being a "single set of premises in single occupation." There are two exceptions to this rule, namely:

(i) where the apparatus comprised in the system is situated in more than one set of premises, but where these are in the same building and are occupied by the licensee or members of his group; and

(ii) where all the apparatus is situated within a single building in single *ownership*. This exception is intended to cover such arrangements as multi-occupied office blocks or shopping centres, where the landlord instals a common PBX or similar apparatus providing services to the residents.

Where under exception (i) the separate sets of premises are not in the same building, then they must be within 200 metres of each other at their closest points, if they are to qualify as part of the same branch system. If this is not the case, each separate set of premises will be individually subject to the BSGL.

Lines (*i.e.* "fixed links" as they are generally called) linking different premises (where not covered by exception (i) above) may not be run by the branch system licensee and must be provided and run by long-line PTOs.

C. SERVICES[2]

1. It might come as a surprise that this particular licence deals at all with the provision of services when it is remembered that its primary aim was to validate what would otherwise have been an illegal act by millions of citizens, the running of a telecommunication system for the benefit of the *user* of that system. In most cases a user would not be conscious of providing a telecommunication service to anyone; it has been suggested that a licensee using a branch system for incoming calls is thereby providing a service to the callers trying to reach him; however, this seems to be over-stretching the arcane distinctions of this aspect of telecommunications licensing.

Nevertheless, within the BSGL limitations on location of the branch system, a licensee can run the system so as to provide services to others (including members of the same group of companies), for example the tenants of a building which he owns.

2. The essential feature of all the simple services permitted by Schedule 3 is that they should extend only to "the conveyance (not including switching) of messages and switching incidental to such conveyance," which obviously encompasses normal telephone service but was thought to preclude the provision of value added (but not basic data conveyance) services. (The meaning of "value added services" for certain regulatory purposes is discussed below).[3] However the Government statement of February 25, 1987, referred to above[4] in relation to BT's licence appears to be of equal applicability to the BSGL. Thus the official line notwithstanding the apparent difficulties posed by the wording of Schedule 3, is that within the ambit of its conditions (particularly those relating to the connection of Private Circuits, contained in Part 5) the provision of certain services (which might be viewed by engineers as value added services) under the BSGL is permitted. One clue, if not a clear justification for this interpretation, may lie in the VADS licence definition of a value added service.[5] If the service merely permits or facilitates the conveyance of the message or its presentation at its destination in "an accurate, reliable and economic manner" it may possibly be treated as part of conveyance or as incidental switching. This could, for example, cover certain electronic mail systems.

D. CONNECTIONS TO OTHER SYSTEMS[6]

1. According to Schedule 3 the systems with which a licensed branch system may lawfully be connected are as follows:

(i) Other branch systems situated on the same premises run by the licensee or on his behalf or by a member of his group.

(ii) Other branch systems in a building occupied by

the licensee and by other persons running such systems provided the building itself is owned by one person. OFTEL gives the example of a landlord or tenant of a large office block with a number of tenants, where a common PBX is provided. The BSGL allows extension apparatus "run" by tenants to be connected to a PBX run by the landlord in such circumstances.

(iii) A Public Telecommunication System.

(iv) A system run by the Crown.

(v) Any other system whose licence authorises it to be connected to the branch system (again this authority would normally be found in Schedule 3 of such a separate licence).

An interesting point of regulatory control arises from these requirements. The permitted connections effectively prevent branch systems being linked directly with other branch systems in order to form a PTO-bypass network.

2. Condition 5, entitled "Connection Arrangements" is one of the most important licence conditions. It contains the basic rule that connection of a system to a public telecommunication system must be by means of network termination and testing apparatus ("NTTA"). The corollary to this is that, under the same condition, except where the connection to the public system is by plug and socket, the licensee cannot himself, or permit anyone else to, make or break the connection of his system with the public system unless so authorised by the public operator.

Where there is a plug and socket connection to the public system, the socket would constitute the NTTA, which according to the BSGL is apparatus on the licensee's premises but comprised in a public system (*i.e.* BT's or Mercury's) enabling the approved apparatus which makes up the branch system to be connected to that public system—in other words to communicate with it. Where the branch system is more sophisticated than a single line system and therefore incorporates Call Routing Apparatus,[7] this would be connected to a test jack frame which itself would incorporate the NTTA owned and controlled by the PTO.

E. USE OF PRIVATE CIRCUITS

Restrictions on the use of private circuits are present in the BSGL in order to limit the potential of private networks to divert traffic and thus revenue away from the public switched networks. The current BSGL represents a relaxation of the previous regime and Oftel is apparently actively considering whether further relaxations can be made, particularly in the light of BT's tariff rebalancing.

1. The provisions as to the use of private circuits in connection with the branch system are set out in Part 5 of the BSGL.

A private circuit in BSGL terms has a specific meaning given in Condition 1(r) which should include a private circuit as described and marketed by a PTO. However PTOs tend not to define Private Circuits in their conditions of contract, preferring to circumscribe their usage by appropriate restrictions, and therefore the BSGL definition is likely to be rather more explicit and self-contained. Its effect is as follows: the "private" element involves conveying messages between fixed points, without switching (which would enable the user to select a destination for the message); and the "circuit" is one or more "communication" facilities (*i.e.* channels) whose capacity is no less than that offered by a PTO at the time of connection of the private circuit to the particular branch system.

For these purposes the definition goes on to provide that a single private circuit will still be regarded as established where, by the use of private (non–PTO) apparatus, individual channels are derived from it; and where multiplexing permits the selective distinguishing and conveyance of messages between different fixed points the communication facility between those points in each case will be regarded as a separate private circuit.

2. The opening condition on private circuits (Condition 12) contains a straightforward ban on Simple Resale, which is defined in the BSGL but in simple terms is the use of a private circuit for carrying a message which has first been conveyed over a public switched system and is subsequently handed back to a public switched system for further conveyance, for example to its destination.

The Government's policy in this regard[8] is explained by its concern that if unlimited use of private circuits were to be authorised, this could jeopardize the cross-subsidy arrangements currently practised by BT for using revenues from profitable main (trunk) traffic routes to subsidise its universal service obligations (BT licence Condition 1). If companies were allowed to use their own networks of circuits leased from PTOs, or offer capacity on such networks to others, this would by-pass the PTO trunk networks and have a dramatic impact on such profitable services, forcing down the revenues of BT which would otherwise be used to provide such cross-subsidy. Accordingly, until 1989 the Government has pledged to retain the simple resale ban, by which time BT's tariff rebalancing policies should be nearly—if not fully—completed. Accordingly, at that stage BT should, possibly in conjunction with sophisticated tariffing of private circuits, be able to cope with such competition without adverse effects to its universal service obligations.

3. There is now no limit on the number of private circuit connections between different branch systems (which could together form a private network) run by members of the same corporate group. (For these purposes a group is a company and all its subsidiaries, "subsidiary" for this purpose having the same meaning as is given in the Companies Act 1985. A group also includes "Specified Former Members," *i.e.* companies who were within the preceding nine months members of the relevant group. However no company can count as a member of two groups at the same time). Given the rather technical drafting of the relevant section of the Companies Act,[9] enabling what could in effect be unconnected companies to be treated as subsidiaries by virtue

of rights of majority Board representation, there may be scope for abuse of the group network exemption. (However, it is understood that in the next Companies Act the general formulation of this definition may be substantially revised.)

Accordingly, there can be unlimited conveyance of public switched network traffic within a private network, provided this traffic is not passed by direct connection, or by a private circuit, to or from any other group's network.

Once traffic goes outside a group network, the rules in Condition 13 begin to bite, depending upon the "kind" of bilateral circuit to which the network is connected. These kinds of bilateral private circuit and their relative restrictions are as follows:

(i) *The first kind*: There is no limit on the number of bilateral private circuits of this kind which may be connected to the group network.
These private circuits are of a kind which does not convey any messages which have been or are to be conveyed either—

(a) by a public switched network; or
(b) by a branch system run by the licensee or a member of the licensee's group, other than such a branch system simply composed of one main system and two "dependent" systems (as defined in the BSGL): an effective limit of two private circuits within the licensee's group network; or
(c) by any other bilateral private circuit.

Generally this rule is intended to cater for external extensions but not inter-PBX circuits.

(ii) *The second kind*: The number of these bilateral circuits that may be connected to any one branch network is limited to three.
This kind of private circuit can convey public switched network traffic (but only one two-way live speech call at a time); however, such traffic is subject to the same restrictions as set out in

(b) and (c) above (*re*, "the first kind") and is also prevented from being conveyed via an IPC.

(iii) *The third kind*: Again, only three bilateral private circuits of this kind may be connected to a branch network.

The description of this particular kind of private circuit is the same as the description for the "first kind," (*e.g.* the messages over it do not go into or come from a public switched network) with the addition of a right to convey messages over that circuit which are going to or coming from another group's system and the further caveat that the private circuit should not convey any message going to or coming from another country via an IPC.

Condition 14 extends the limitations in Conditions 12 and 13 to cover international private circuits as well.

5. Where a branch system is connected to a second, non-BSGL system, either directly or via a private circuit, and the person running that second system has notified the branch system licensee of the application of certain conditions set out in the "call-up conditions" in Annex B to the BSGL, the licensee is bound by those conditions in relation to messages passing between the licensee's branch system and that second system. Accordingly, a licensee under the BSGL connected to a non-BSGL system should be told by the licensee of that system which conditions in Annex B apply to connection of the two systems. These conditions will cover both PSTN and non-PSTN traffic and may replace or be in addition to the connections allowed under Condition 13.

Annex B appears to be an attempt to deal with the perhaps unforeseen problem that there would be other licences besides the BSGL which would contain different "hop" rules. Annex B is the mechanism for solving this problem for class licences (such as VADS) and also for individual licences which are modelled on such class licences.

F. MAINTENANCE SERVICES AND DESIGNATED MAINTAINERS

The BSGL conditions in connection with maintenance apply only to Call Routing Apparatus and are to be found in Part 4 of the BSGL, Conditions 9 to 11 inclusive.

"Call Routing Apparatus" is carefully defined in the BSGL as meaning apparatus installed *and connected* to be capable of switching two-way live speech calls between at least two internal extensions and at least two exchange lines.

(a) Maintenance Services

"Maintenance Services" has a wider meaning under the BSGL than would normally be given to maintenance. Such services include not only carrying out repairs and verifying that the apparatus is properly performing to specification or complies with a condition contained in any approval under section 22 of the Act, but also extend to an activity involving the removal of the outer cover of the apparatus, including alteration of any stored commands affecting the compliance of the apparatus with technical requirements of public operators.

General Condition 9 essentially requires that any maintenance services to be carried out on call routing apparatus should be undertaken by a Designated Maintainer. This is a person who in relation to such apparatus is either the operator of the public system to which it is to be connected or otherwise is an approved contractor (under section 20 of the Act). Unless the Designated Maintainer is itself running the licensed system it must have a contract with that person for the provision of the Maintenance Services.

Designated Maintainers are approved by the Director General after certification by the British Standards Institute, which operates a registration system for maintainers in accordance with BS 5750, Part 2. However, BT and Mercury are not required to register with BSI as Designated Maintainers and strictly speaking are thus not subject to same requirements.

These rules as to maintenance of call routing apparatus also apply to wiring connected to it and any extension telephone apparatus which is hard wired to it. Where the operation involves the removal of the outer cover of the call routing apparatus or involves wiring connected to it, exceptionally the licensee may allow inspection of the apparatus or wiring for assessment as to whether it is fit to be maintained by the person making the inspection; for this the Designated Maintainer's permission should be obtained or the Designated Maintainer should have failed to give that permission and a period of 14 days should have elapsed.

A licensee may also allow a public system operator to test the operation of that public system or to "Bring into Service" any apparatus to be connected to that public system. Presumably the facility for inspection is to enable a licensee to make the necessary arrangements for a change in his Designated Maintainer, which might have been frustrated if the intended maintainer was not able to inspect the apparatus beforehand.

(b) Entry into Contracts for Provision of Maintenance Services

General Condition 10 requires the licensee to have any one Designated Maintainer and to give him all necessary access to the call routing apparatus and its wiring. The licensee is also required not to terminate the contract of any Designated Maintainer except by at least 90 days' notice in writing or on repudiation of the contract by the Designated Maintainer.

(c) Bringing Apparatus into Service

"Bringing into Service" also has its own special and somewhat extended meaning, set out in the BSGL. Primarily it means the process of connecting or disconnecting telecommunication apparatus to or from a telecommunication system but it also includes testing and inspection in order to ensure the apparatus, or the system in which it is comprised, is authorised to be connected to another system.

Condition 11 requires that any item of apparatus which is to be comprised in a branch system and is *not itself directly* connected to a public system, and which is connected to call routing apparatus, can only be brought into service by the Designated Maintainer of that apparatus or "any other person" where the Designated Maintainer has agreed to that other person carrying out the work or has failed to bring the apparatus into service before the end of 14 days after notice from the licensee requiring the apparatus to be brought into service. In this context "any other person" appears to mean literally that, as such person would not have to be an approved contractor.

(d) Maintenance Contracts

The above provisions as to maintenance services and bringing into service contained in the BSGL merit comparison with PTOs' and Designated Maintainers' actual contracts for the provision of service. In recent times certain of BT's published contracts for such service appear to have been inconsistent with some conditions of the BSGL. For example, they do not or did not oblige BT to provide maintenance service within a fixed period. Termination provisions have also not necessarily been consistent with the 90 day rule in the BSGL. Worse still, where a *force majeure* or some other situation arises which will manifestly prevent the Designated Maintainer from bringing apparatus into service nonetheless the licensee must go through the formality of giving notice to the Designated Maintainer and waiting 14 days before any other person can carry out the necessary work.

Furthermore, the licensee seeking to exercise his rights, under Conditions 9 and 11, to allow persons other than the Designated Maintainer to carry out necessary work, would also have to be assured that such action would not put him in breach of his maintenance contract with the Designated Maintainer.

Up until the time of writing such maintenance contracts were not written in terms reflecting these provisions of the BSGL and indeed went so far as to oblige the licensee not to allow any third party to tamper with the apparatus under any circumstances. This is understandable in a purely contractual context but obviously poses potential conflicts with the BSGL regime, not to mention confusion to customers.

Note also that under Condition 9 there is no time limit afforded to the licensee for compliance by the Designated Maintainer, after which another approved contractor could carry out the work; a somewhat surprising loophole.

G. Other Obligations on BGSL Licensees

The remaining obligations essentially concern the keeping of records, use of wiring and technical requirements, as follows:

(a) Requirement to Furnish Information to the Director (Condition 1)

This requires a licensee to provide the Director General on request with documents and other information which he may reasonably require (and without placing an undue burden on the licensee) in order to exercise his functions under the Act. This would be, for example, so that the Director General could assure himself the licensee was running the system within the scope and terms of the licence grant.

The Director General is also entitled to inspect the licensed system for the same purposes and to ensure that the connection of any other system to it would not cause any contravention of the licence applicable to that system.

In addition, for a system comprising call routing apparatus and first connected to a public system after July 1, 1983, or connected to a private circuit, the licensee is also to maintain a record containing certain information set out in Annex 1 to the General Conditions. This information includes basic details as to the identity of the licensee, his public systems connections and information on call routing apparatus and private circuits to which the licensed system is connected.

(b) Wiring etc Forming Part of the Applicable System (Condition 2)

This particular condition is designed to free-up the use of internal wiring under the control of persons running telecommunication systems who, generally speaking, are not using the particular wiring for the purpose of running their own particular systems. This would apply, for example, to a landlord whose building contains wiring which he owns but which serves various tenants in that building.

Under this condition the licensee must make apparatus, for example, wiring, available to anyone wishing to use it in the running of any telecommunication system ("the user") either by renting it out upon reasonable terms or by selling for a reasonable lump sum. There are safeguards to ensure that the terms offered by the licensee are fair and comparable to those that would apply if the licensee actually ran the other party's system or provided maintenance services for it or supplied apparatus comprised within it.

In addition the licensee is bound to allow the user to carry out operations on such apparatus and/or controlled by the licensee to enable the user to use it in order to run his telecommunication system. An exception is made to this principle where the apparatus, again typically wiring, is installed with other apparatus in such a way that Maintenance Services cannot be carried out on that apparatus without disturbing or affecting the other apparatus, for example where both such apparatus is installed in a shared casing.

It should be noted that these requirements on the licensee only apply to apparatus (*e.g.* wiring) which the licensee owns and which is installed on premises occupied by the user and which either was, but is no longer, comprised in a telecommunication system run by the licensee or continues to be so comprised but where the user "is not obliged to allow the licensee to continue to run" such system.

(c) Compliance with European Community Requirements (Condition 3)

This Condition requires the licensee to comply with "Relevant Community Requirements," defined as being requirements relating to specifications, functioning or use of the licence system in order to secure compliance with a Community obligation and which is described in a list kept for the purpose by the Director General and made available for inspection by the general public.

These Community obligations are in fact being drawn up by the CEPT and their introduction will also serve to implement European Community Directive 86/361/EEC on the initial stage of the mutual recognition of type approval for telecommunications terminals. These standards, known as NETs (Normes Europeennes de Telecommunications) will supersede any prior national standards within their scope of application.

Some harmonisation of requirements in the area of electrical safety is being studied by CENELEC (European Committee for Electro-Technical Standardisation). The Commission of the European Community has also proposed a directive laying down standards for electromagnetic compatibility for all kinds of equipment.

(d) Technical Requirements (Condition 4)

This provides that if the system comprises more than one item of apparatus and is connected to a public telecommunication system then, unless the connection is merely temporary in order to commission call routing apparatus, the system must comply with technical requirements laid down by the Director General.

It is pursuant to this condition that the NCOP has been issued.[10]

(e) Relevant Operations on Apparatus Comprised in Specified Telecommunication Systems (Condition 6)

Under this provision any public system apparatus installed on the licensee's premises cannot be tampered with by the licensee nor is the licensee allowed to permit anyone else to tamper with such apparatus without the prior authorisation of the public system operator.

(f) Connection of Automatic Calling Equipment to Specified Public Telecommunication Systems (Condition 7)

This provision polices what might be called "junk" telephone calls. Automatic calls, not involving live speech, but having recorded or synthesised speech or other sounds, sent by automatic calling apparatus, may not be used to send such messages to anyone unless that person has consented in writing to receive a message. The licensee is bound to maintain a register of the persons who have consented to receiving such messages, which must be made available for inspection on reasonable notice by the Director General.

(g) Emergency Telephones for the Hearing Impaired (Condition 8)

If the system is installed in a lift to which the public have access, the licensee is bound to take reasonable steps to ensure that a telephone installed in such lift can be coupled inductively to hearing aids designed to be so coupled to such a telephone.

Notes

[1] Annex A to the BSGL.
[2] Sched. 3 to the BSGL.
[3] See Chap. 6 at pp. 102, 103.
[4] See Chap. 4 at p. 48 above.
[5] See n. 3 above.
[6] See n. 2 above.
[7] Discussed below at p. 89.
[8] See also Chap. 1 at pp. 11, 12.
[9] Companies Act 1985, s.736.
[10] See Chap. 8 below at p. 131.

Chapter 6

VALUE ADDED AND DATA SERVICES LICENCE

A. Background

1. As noted in the previous chapter, the BSGL, insofar as it permits the provision by licensees under that licence of services to others, purports to limit these services to the conveyance of messages and switching incidental to such conveyance. Whilst this description is no longer treated as precluding certain value added services, the BSGL, as has been seen in Chapter 5, contains certain restrictions, notably on the use of private circuits, which could be inhibiting, particularly for international VADS networks. Undoubtedly the BSGL's scope would, as mentioned in Chapter 5, extend to conveyance of data, although a licensee providing such a service under the BSGL would again be subject to its routing constraints.

Value added services are essentially computer-controlled operations involving such things as message handling, including store and forward, network services, such as protocol conversion, electronic mail and similar services. No network as such is necessary in order to provide such services, which can simply be operated by means of a computer attached to the end of a public telecommunication system.

By contrast, data services in the nature simply of conveyance of data can only be provided over lines, *i.e.* circuits or channel capacity within circuits, made available by PTOs. Where such services are made available by the lessees of such circuits directly to others this is resale, but should not be confused with "simple resale," prevented under Condition 2.1(a).[1]

2. Until the end of April this year, value added services could only be provided pursuant to the VANS Licence issued under the 1981 Act. As VANS services grew as an industry and became more sophisticated, the operation of managed data networks emerged and the Government, following the introduction of the Act, embarked on an extensive consultation exercise over its proposed Class Licence for value added services and managed data networks. Eventually, this exercise must have convinced the Government that if the rules were to be relatively simple to follow and to supervise, it would be more expedient to license the provision by all-comers of everything but basic service or, in other words, "basic conveyance," leaving voice telephony (live speech) and basic telex as the preserve of the PTOs.

Gradually the term "value added services" seems likely to disappear from telecommunications parlance as rapidly as it emerged in the early 1980s, as ISDN develops and these services became increasingly integrated with basic network services.

The DTI has issued a useful explanatory guide to the VADS licence.[2]

B. LICENCE GRANT

1. The VADS Licence runs for a minimum 12 year period (with provision for continuation subject to revocation on two years' notice) and applies to anyone other than public telecommunication operators (for example BT, Mercury and Hull) and their Associates. PTOs and PTO Associates are separately licensed for such purpose.[3] VADS licensees are referred to generally in this Chapter as "Service Providers."

2. The licensed systems are described in Annex A as "telecommunication systems of every description within the United Kingdom," in precisely the same terms as the licensed systems described in the BSGL. Accordingly the apparatus comprised in the VADS system and run by the Service Provider must be situated within a single set of premises in single occupation or otherwise be situated in different sets of premises occupied by the same person or by members of the same corporate group (*i.e.* parent company and its Companies Act subsidiaries), provided the different sets of premises are within the same building or are within 200 metres of each other (the "200 metre rule"), or are situated within a single building in single ownership.

Although therefore the VADS licence liberalises the provision of data and value added services, outside of the 200 metre rule the fixed links through which such services are provided, whether by cable, microwave or other transmission medium, still therefore require to be installed and run by PTOs.

3. In a private network of "nodes" linked together by PTO-provided private circuits each such node would itself constitute a system able to qualify as one licensed under the VADS licence (or any other appropriate licence, *e.g.* the BSGL). The private circuit links would of course remain part of the systems run by the particular PTO under its own PTO licence.

4. It should be remembered that where the Service Provider's system is liable to fall outside the Scope of the VADS licence, perhaps because of difficulty over strict compliance with Schedule 3 or Annex A or because some of the conditions in Schedule 1 are inapt, there remains the possibility of an application to the DTI for an individual licence. Obviously the authorities will wish to be given convincing reasons as to why a person should be singled out for individual treatment in this way; also in such circumstances a special licence fee would be payable.

C. LICENSED SERVICES

1. Schedule 3 contains the authorisation to connect the VADS licensed system to other systems as listed. This includes "Specified Public Telecommunications Systems," which are systems designated by the Director General for the purposes of the VADS licence specifically. Different public systems could be so designated under different licences.

Schedule 3 also contains, in paragraph 1(*b*), the authorisation to provide the telecommunication services described. In describing such services the DTI have abandoned any attempt to define "Value Added Services." Instead, all forms of telecommunication services provided by the licensed systems are authorised except cable programme services, land mobile radio services and services provided to persons *outside the licensee's Group* the only *substantial element* of which is conveyance of live speech or telex messages. Where the value added service itself is substantial in relation to the whole service (not simply telecommunication service if it also comprises, say, a computer service) application of the restriction on conveyance of live speech or telex will obviously not arise.

At first sight it might seem debatable whether a Service Provider could be held to "convey" anything, given that the Service Provider is reliant upon a PTO for taking a message to its destination. However it must be remembered that by virtue of section 4(7) of the Act "conveyance" has an expanded meaning and includes transmission and switching (as well as reception of messages).

The prohibition on live speech or telex service only applies where the service is provided for "a consideration." The DTI view is that for these purposes 'consideration' should bear the same meaning as at Common Law. However, whilst it is clear that any nominal consideration such as a peppercorn will suffice it is less clear whether this will include those operators of in-house networks who provide services "free of charge" to customers, suppliers, etc. It is arguable and probably safest to assume in such cases that the broad benefits derived by the Service Provider from its ordinary business relationship with such persons provide some form of indirect consideration for the services provided by the Service Provider.

2. In considering the live speech/telex restriction there remains the question of what is meant by the "only substantial element." Dictionary definitions of "substantial" tend to give alternatives, such as "considerable," and are therefore of little help. There is also the technical difficulty of how "substantial" is to be judged, whether by reference to price, relative proportions of different services (telecommunication and non-telecommunication), capacity or bandwidth available or otherwise.

In measuring this "substantiality" it is therefore important to determine the factual context. The licensed system may be providing non-telecommunication as well as telecommunication services as the latter are defined in section 4(3) of the Act. The use of the simple word "service" in the context of the telecommunication service forming the whole or part of a "service" suggests the wider context of both types (telecommunication and non-telecommunication) of service and therefore greater latitude for live speech or telex. For example, this would have tariffing implications in that under Condition 13 of the licence the Service Provider is bound only to publish and notify the Director General of the prices for *telecommunication* services

It should be noted that the licence conditions refer in many places[4] to a Relevant Service, defined as *any service* provided by the licensed system.

It almost goes without saying that the restriction on live speech and telex will not extend to recorded speech or telex reforwarding, provided the storage of the messages in those instances is more than simply temporary and incidental to their conveyance.

D. LICENCE CONDITIONS

1. *Condition 1* set out in Schedule 1 to the licence brings into play the General Conditions applicable to all Class Licences. These Conditions are described in some detail in Chapter 5.[5]

2. *Condition 2.1(a)* prevents the provision by the Service Provider of a Simple Resale Service. Simply put, this covers the situation where a message is conveyed from the public switched network, over a leased line and then back onto a public switched network without any additional service or facility being provided in respect of such conveyance. The ban on simple resale, to which the Government is committed until at least 1989, is discussed in more detail in Chapter 4.[6]

3. *Condition 2.1(b)* prohibits the Service Provider providing a service for reward in which sounds are conveyed (alone or together with visual images), and the only function performed by the licensed system is multiplexing together with "transmission and reception." It would appear this was intended to prevent the basic conveyance of sound and associated vision by means of splitting larger capacity circuits. However, according to the DTI it does not prevent video conferencing services or other services such as broadcast network facilities management, in which the multiplexing may be an incidental part of the comprehensive service.

4. *Condition 2.2* covers the situation in which live speech services to persons outside the Service Provider's Group are provided for reward otherwise than as a substantial element of a Relevant Service. If live speech were the only substantial element it would, of course, fall foul of the restriction in Schedule 3. Where live speech is thus legitimately provided as an insubstantial element, the Service Provider must levy a charge which is either 25 per cent. above the lowest charge which would have been levied by a PTO or equal to the highest charge leviable by a PTO, whichever is the lesser. This raises the problem that there is no requirement that separate charges for different services be published by the Service Provider and charged to the Customer. In the absence of such a requirement it may be difficult to police this provision as the Service Provider will be free to "bundle" charges for ancillary voice services with other services.

5. *Condition 3* applies where the Service Provider is providing services for no consideration. Typically, this will be in the case of an in-house or closed user group network, with the Service Provider's "customers" able to use the network free of charge. In such a situation, assuming there is no question of "consideration" being present,[7] the Service Provider is free to provide any service, including voice and telex. In the case of data traffic where the Service Provider is making a charge for the service these restrictions would obviously not apply. However the fair trading conditions and certain of the OSI provisions[8] would then be brought into play.

Reflecting the equivalent BSGL provisions on Private Circuits, Condition 3.2 provides that whatever the service provided, no message can be conveyed over more than one private circuit which links the Service Provider to a different Group (*i.e.* a "Bilateral Private Circuit"). This does not prevent the Service Provider from being connected by any number of Bilateral Private Circuits to several different Group(s) "A", "B", "C" and so on but it does prevent a message being conveyed via a "further hop" from, for example, Group A to Group B (or from Group A to Group C via the Service Provider).

6. Restrictions in this Condition, in relation to live speech and telex, on the number and type of connections that may be made to the Service Provider's System only apply where Private Circuits link that System with a person outside the Service Provider's Group. These restrictions therefore do *not* apply where the Service Provider's network is serviced over a PTO's Public Switched Network.

Again, mirroring the BSGL, Conditions 3.3 and 3.4 limit the numbers of such private circuits that may be connected to the Service Provider where voice or telex is involved and also the numbers of "hops" that may be made.[9]

Further guidance, with useful diagrams, is given on the Condition 3 routing constraints in the DTI's Explanatory Guide on the VADS licence.[10]

7. *Condition 4* deals with the use of International Private Circuits. The old VANS Licence was defective in that it did not deal adequately with the use of IPCs for the provision of VANS. Condition 4 is an attempt to improve the position as much as the DTI can in the light of CCITT Recommendations. The basic principle that has been adopted is that where a service is provided for reward, Value Added Services are wholly liberalised, whereas Data Services will be permitted only if provided to a "Closed User Group." There are however no such restrictions where no "consideration" passes (for a discussion of the scope of this term see above).[11]

This is the only condition in the Licence where the distinction between Value Added Services and Basic Conveyance has significance. "Value Added Service" is defined as any Relevant Service[12] where there is a service additional to Basic Conveyance (also defined). It therefore covers any deliberate removal of or addition to the information content of the message or any other additional services except those which merely permit or facilitate the conveyance of the message or its presentation at its destination "in an accurate, reliable and economical manner."

If not a Value Added Service, to legitimise the provision of a Relevant Service over an IPC, a Closed User Group must be involved, *i.e.* any person or group of persons having a common business interest other than in the provision or securing of telecommunication services. This is an adaptation of the comparable definition used in CCITT Recommendation D.1 paragraph 6. Any such Closed User Group must be approved by the Secretary of State and registered in a list kept by OFTEL, available for general inspection.

There is also in Condition 4.2(iii)(b), one limited exception to the prohibition on the provision of basic data services over international private circuits. This applies to any service which is specified by the Secretary of State, pursuant to any international agreement between the United Kingdom and a foreign telecommunications administration. The DTI announced recently its intention to pursue such agreements with the United States and with Japan.

Assuming that the Licensee is providing a Value Added Service or provides Basic Conveyance to a Closed User Group, there is still the problem in Condition 4.1 that no "International Simple Resale Service" can be provided. As with the definition of "Simple Resale Service" no consideration need necessarily be involved. The definition of this term covers the situation where a call is taken from a United Kingdom public switched network, conveyed over an IPC, and then in turn is conveyed over the public switched system of the foreign PTT.

8. *Condition 5*: imposes a requirement on the Service Provider to give a special notice to any customer outside the Service Provider's Group and with whom the Service

Provider's System is connected, either directly or by means of a private circuit. The notice must specify the "Call-up" conditions as to routing which are set out in Annex B to the customer's licence and which the customer must observe in conveying messages which have been conveyed over the Service Provider's system. These rules, as to live speech and basic telex, are the same as in the BSGL. There are no routing restrictions (other than for simple resale[13]) on value added or data service traffic provided by a Service Provider for a consideration. Any private circuit connections between the Service Provider and its customer are additional to the "two-hop" bilateral private circuits allowed under the BSGL. Note that Condition 5 does *not* apply where the Service Provider and the customer are connected over a public switched network and not by way of a Private Circuit.

9. *Condition 6* prohibits linked sales or tie-ins in much the same terms as BT's licence equivalent.[14] Condition 6.3 however permits the offering of quantity discounts and the bundling of equipment and services where "economies of scope" are involved.

10. *Condition 7* obliges the Service Provider to safeguard the privacy of its customers' messages.

11. *Condition 8* on Numbering Arrangements, requires Service Providers to adopt suitable addressing and numbering arrangements in a Numbering Plan to be filed with the Director General within three months of starting the service. However, if such Numbering Plan is found to be incompatible with any other operator's numbering scheme filed with the Director, or does not allow for sufficient numbers to be available, the Service Provider may have to change the Numbering Plan. Therefore before full adoption and implementation of a Numbering Plan it will be as well to consult as closely as possible with OFTEL. OFTEL have prepared a special guidance note on "The Numbering Arrangements for Value Added & Data Services."[15]

E. FAIR TRADING CONDITIONS

1. The VADS Licence contains a set of fair trading conditions along very similar lines to those included in PTO licences. (These conditions are therefore also discussed in relation to BT.[16]) Given that licensees under the VADS Licence will compete with such PTOs offering value added services and vice versa, it was thought appropriate that for the purposes of value added and data services substantial licensees should operate in the same competitive environment. These licensees have been dubbed "Major Service Providers" by the VADS Licence. In order to be categorised as a Major Service Provider the licensee would have, in the financial year immediately preceding the year in question, a group turnover exceeding £1 million in respect of value added and data services or a group turnover exceeding £50 million from sources of any kind.

These fair trading rules also apply to providers of "Trilateral Services," being operators of systems which pass messages between at least two other systems operated by parties not associated with the VADS operator.

VADS are not subject to the Fair Trading and OSI Conditions where the services are only provided to persons within the Service Provider's Group.

2. The main provisions of note for Major Service Providers are as follows:

(a) Undue Preference/Undue Discrimination (Condition 10)

The Service Provider must not show undue discrimination or undue preference to any person in the provision of services. This applies not just to third parties but to members of the Service Provider's own corporate group, who must, therefore, obtain services at standard tariff.

Generally this provision is very similar to the equivalent prohibition contained in the main PTOs' Licences. The major difference is that the Director General is not given any express power to determine any question of what is to be regarded as undue preference or undue discrimination. (It is arguable that in class licences such a discretion cannot be given to the Director General[17]). However, the Director General has a duty to ensure compliance with the Licence conditions and in that event it would first fall to him to decide the issue in determining whether or not a compliance order should be made under the Act.

(b) Prohibition of Cross-Subsidies

The Service Provider must not *unfairly* cross-subsidise the value added and data services from any other source, where these services are provided for a consideration outside the Service Provider's Group.

(c) Separate Accounts (Condition 12)

Separate accounts must be kept for a Relevant Services Business by no later than April 1, 1988 or within 12 months of first running the licensed system, whichever is the earlier date.

(d) Publication of charges, terms and conditions (Condition 13)

The Service Provider must publish either specific charges or a method of determining charges, and the terms and conditions on which services are offered within 28 days after the service is first provided. The Service Provider must not depart from such charges (or method) or such terms and conditions. One day's notice must be given to the Director General of any changes. Note that this obligation only covers that part of the service which involves the conveyance of messages and/or the connection or installation of apparatus. Publication is by way of filing the relevant information with OFTEL and by making available the information in a publicly accessible part of the Service Provider's major office.

Note that the Director General may give a written consent dispensing with the requirements of Condition 13 in individual cases. This is only likely to be possible where in practice a service is not being provided or offered to the public generally.

(e) Customer Confidentiality (Condition 16)

The Service Provider must introduce a Code of Practice on the confidentiality of customer information, and must ensure that its employees observe such code. The Service Provider must submit its code to the Director General for his approval within three months of starting to run its system. Alternatively, the Service Provider should adopt OFTEL's published model code.

(f) OSI Standards (Conditions 17 and 18)

On the technical side, the Government has for some time accepted the argument (although at present the industry would not necessarily agree) for expanding the use of OSI standards in networking and accordingly the VADS Licence provides that once an OSI standard has been specified by the Director General a licensee who is a Major Service Provider must ensure within 12 months from that time that means of access to the service are capable of being provided by OSI standards. Advance notice must also be given to the Director General of proposed changes to such means of access where such changes might cause other Service Providers or PTOs or suppliers of apparatus materially to modify, replace or cease to produce or supply any item of apparatus which was to have been connected to the VADS system.

These obligations to provide access to Services by means of OSI standards (which potentially are quite onerous and could involve additional—sometimes unbudgeted—expenditure by Service Providers) do not apply where the Service Provider is unable to obtain the necessary equipment or facilities to permit the connection to be made in this way or where a customer who would otherwise be entitled to OSI facilities requests these when the necessary apparatus has not been installed or completed or such apparatus has not been adapted or modified as necessary. The VADS Licence includes fail-safe mechanisms for ensuring that when feasible these facilities will, however, be provided, in accordance with time limits agreed with the Director General.

OFTEL have published notes[18] explaining how the Director General intends to consult before specifying a particular standard and how conformance to standards can be assured.

(g) Fees (Condition 19)

A Major Service Provider or Provider of Trilateral Services must notify OFTEL within 30 days of the date on which the licence conditions first apply to it of its name, address and Major Office, normally its registered office. Major Service Providers must pay an annual fee within 30 days of such notification and annually thereafter, such fee to be determined by the Director General. At present such fee is £1,000.

F. PUBLIC TELECOMMUNICATIONS OPERATORS AND THEIR GROUP ASSOCIATES

1. As explained above, the VADS Licence excludes as licensees public telecommunications operators and their Associates.

Value added and data services offered by PTOs such as BT and Mercury fall under the licences granted to them in 1984, *i.e.* their general PTO licences. Their Associates are covered by a further Class Licence issued on the same day as the VADS licence.

For these purposes an Associate means not only a company which is a member of a PTO's group (for these purposes a group comprising a holding company and all its subsidiaries, within the meaning of those expressions given in section 736 Companies Act 1985) but also a company in which a member of a PTO's group has the right to more than 20 per cent. of the votes or more than 20 per cent. of the profits, as well as a subsidiary of any such company.

2. The main reason why the Government wished to have a separate licence for PTO Associates was to give it some kind of vetting power over potentially powerful groupings of PTOs and other substantial companies, particularly from the computer industry. This has been translated into a particular power of revocation available to the Secretary of State and set out in paragraph 1(*c*) of Schedule 2 to the Group Associates licence.

This provides that where the licensee notifies the Director General (which it must do under Condition 20.3 of the licence) that it has become an Associate of a "Relevant Company" (see below) or that a Relevant Company of which it is already an Associate has begun to provide Relevant Services and within 40 days of such notice the Secretary of State decides that competition in the provision of Relevant Services is likely to be "adversely affected," the application of this class licence in relation to the particular Associate may be revoked on 30 days notice. There is a similar revocation power for the Secretary of State where a company of which the licensee is an Associate has been involved in a transaction which has resulted in two or more businesses coming under common control, following which that company is a Relevant Company.

3. For these purposes a Relevant Company is defined as being as a company whose turnover (where separately identifiable in the Company's accounts, in computer or telecommunication products or services) exceeded £1,000,000,000 or a higher figure determined by the Director General in respect of all relevant companies and their groups in the latest financial year. Relevant Service means any service provided by the licensed system, *i.e.* a value added or data service.

In paragraph 1(c)(i)(bb) the word "it" appears to be missing in the second line after "relevant company". Nonetheless the intention of this provision would appear to be to catch PTO Associates which are already linked to Relevant Companies and also PTO Associates which at the time of their associating with a Relevant Company are not providing value added or data services but do so later at which point the Director General might wish to review the situation.

Notes

[1] See p. 100 below.
[2] Explanatory guide to the Telecommunications Class Licence for the running of systems for value added services.
[3] See p. 108 below.
[4] See for example para. D.4 below.
[5] At p. 80 above.
[6] At p. 78 above.
[7] See p. 99 above.
[8] At p. 107 below.
[9] For further discussion of these rules see Chap. 5 above, at pp. 87 *et seq*.
[10] See n. 2 above.
[11] See n. 7.
[12] See p. 100 above.
[13] See p. 86 above.
[14] See p. 64 above.
[15] OFTEL 5.5.87.
[16] See Chap. 4 above at p. 59.
[17] See s.7(6) of the Act.
[18] OFTEL 27.8.86.

CHAPTER 7

TELECOMMUNICATION SERVICES

A. VOICE

1. The (Conservative) Government's so-called "duopoly policy," as it has become known, is currently intended to remain in force until at least November 1990. This stems from Mr Kenneth Baker's statement in November 1983,[1] but although the Government has widened competition, particularly in the field of data communication (see the VADS Class Licence[2]) it is currently not prepared to allow open competition in the provision of voice or basic telex services other than between BT and Mercury.

Accordingly, the provision of voice telephony remains completely under the control of BT and Mercury (and, within its own area, Hull Council) with the exception of cellular radio and mobile radio telephone licensees, cable television licensees and (for the limited purposes) Electronic Data Systems Limited ("EDS"). Currently, the licences granted to cable television operators permit the provision of voice telephony services by means of their cable systems so long as this is done in conjunction with either BT or Mercury. This means in fact that a cable television licensee must ensure that the part of its system used wholly or partly for providing voice telephony service is run and maintained either by BT or Mercury or, somewhat contradictorily, by the licensee itself as agent for BT or Mercury. This retains at least a semblance of control over the voice telephony operation for BT and Mercury. Accordingly, it is necessary for the cable television licensee to reach an agreement with BT or Mercury before such services are provided. This has provoked criticism from certain cable TV licensees.

2. Until recently, voice messages were transmitted simply to and from customer terminal apparatus, such as a PABX, through exchanges, local and trunk (in BT's system these are now being replaced by a new digital hierarchy of exchanges) which switched these messages (unless a leased or private service was being provided) to and from their destination. Now a new type of facility for voice telephony in particular is beginning to be offered by the PTOs, Centrex. As the name suggests, this is a service which removes the need for a customer to have a sophisticated PABX on his premises, as all the switching and incidental functions are carried out by equipment within the PTO's system.

Typically, a part of the PTO's digital switching unit or exchange would be dedicated to providing the Centrex service and this would include enhanced facilities such as "follow me," redial and call back, conference calls and "camp on" which are to be found in modern PABXs. The consequential creation of a private switching facility for the customer has thrown up new and completely unanticipated regulatory problems, for example on appropriate standards and, in the case of PTO licences, in relation to the definition of "simple resale."

This latter problem arises out of the use of dedicated circuits, which would be deemed to be private circuits under these licences, in order to transmit messages between the various premises of a Centrex customer; under licensing regulations (the BSGL) these "private" lines combined with the treatment of lines from the Centrex switch to the customer's apparatus as service lines would have resulted in the arrangements amounting to simple resale. As this is prohibited by PTO licences (Condition 46.1 of BT's Licence) OFTEL decided to overcome this anomaly by a modification to BT and Mercury's licences which has now been made. For this purpose the definition, in BT's licence Condition 46 (Private Circuits) and in the Mercury licence equivalent, of "Simple Resale Services" has been completely revamped. The effect is to leave out of Simple Resale a conveyance of a message for which the charge is a "Relevant Tariff," for example a special tariff for the portion of the PTO's system providing the Centrex service.

3. There are limited exceptions to the restriction of voice services to BT and Mercury, appearing in the

Schedule 3 to the VADS Licence. Their effect is that live speech (as well as basic telex) services may be provided by any VADS licensee to members of its own corporate group, for "a consideration." Live speech services may also be provided to any other person, again for a consideration, provided these services are only a small part (*i.e.* not "the only substantial element") of the full service provided to the customer. The EDS licence is not quite so restrictive.

Note also that as the current regulatory restrictions relate to "live speech" they would therefore not extend for example to the provision of recorded messages. Where these are provided in traditional form (for example the Speaking Clock) such services should be authorised by the BSGL but where recorded voice messages are enhanced in any way (for example voice mail) this would normally fall to be covered by the VADS Licence.

4. The individual licence granted to EDS on September 30, 1987 is also a potentially significant, albeit unique, exception to the otherwise monopoly of BT and Mercury in providing voice telephony. EDS appears to be the only non-PTO so far licensed to offer voice services to more than one party via a common network. Its licence, for a managed network, permits the "layered" (or partitioned) use of switching and PTO provided transmission facilities to provide "managed" voice (live speech) services and integrated voice/data services. The partitioning of the EDS service arises from restrictions in its licence preventing EDS conveying voice or telex messages between its various customers, so that such conveyance is permitted only within one defined corporate group at any time.

B. DATA

1. The expression "data" covers a broad spectrum incorporating, for example telex, facsimile, electronic mail, packet switching and video.

Message handling systems using the CCITT X.400 subset of Open Systems Interconnection will soon bring together the different electronic mail systems and allow customers of different service providers to communicate with each other.

2. Telex has been perhaps the most popular medium for data communication for several decades and although telex connections are still growing (at about 7 per cent. annually) telex is being gradually overhauled by facsimile and eventually electronic mail seems set to surpass both as it will probably be cheaper, faster and perhaps less error-prone. One of the reasons for the proliferation of telex is that it has at present the unique feature of an answerback which provides evidence of receipt. Moreover, where security is required, messages can be encrypted and can also be authenticated using a "test key." This is an algorithm agreed between the correspondent parties using random table numbers for the date and serial number of the message and other information contained within it.

Telex is defined by the CCITT (recommendation F.60) as "a telegraph services for subscribers, whereby they can communicate directly and temporarily between themselves using start-stop telegraph equipment operating at 50 bauds and with International Telegraph Alphabet no. 2." The VADS Class licence also contains an expanded definition of Telex, since a service "the only substantial element of which is conveyance of live speech or Telex messages" is excluded from the telecommunication services authorised by that licence.

3. Facsimile messages currently utilising Group 1, 2 and 3 terminals, are for now conveyed over the public switched networks. However, Group 4 terminals will be able to work over digital transmission paths.

4. Packet switching was developed for communication between computers. The information being sent from one computer to another does not travel in a continuous stream and each computer is connected to a node which sub-divides messages from the computer into a series of data "packets." Each message contains information sufficient to identify its routing and the network acts rather like a postal service in reading this information and delivering the packet within a fraction of a second. There is no "full time" connection between the sending and receiving computers and therefore considerable time as well as capacity is saved through not having to open up a circuit between them. This technique enables the network to switch and convey packets constantly, filling the time gaps between packets going to one destination with packets going to another.

In packet switching the receiving node is normally programmed to reassemble the packets in proper order before delivering the message which they together comprise.

Packet switching may sound like a value added service but in regulatory terms (*e.g.* PTO licences) is to be treated as a basic telecommunication service. Accordingly, this type of data service falls within a PTO's "Systems Business' and would be caught by the PTO's licence obligations to provide basic service as well as to interconnect the relevant parts of its system to other public operators' systems.

5. The operation of systems and networks for the provision of data services, whether or not there is any value added element, requires a licence under the Act. In most cases either the BSGL or the VADS Licence will apply, unless their terms[3] cannot be met.

There are no regulatory constraints over the way in which data services may be provided other than those appearing in the applicable licence. Accordingly, the service may reach the customer by satellite (for example point to multi-point satellite service) or by wireless telegraphy (for example mobile radio/cellular radio and radio teletext) or by terrestrial links, but so far (though the Government is reviewing the position as regards services by satellite[4]) the physical provision of such links remains the preserve of BT and Mercury nationally (and of cable television PTOs in their localities) and generally capacity must be leased from BT and Mercury in order for a third party to provide such service to its customers.

6. Radio and television teletext transmissions, embodied in a "sub-carrier" of the BBC or IBA broadcast signal, are increasingly being used to provide data services. Whilst authorisation for such transmissions will be required from the Cable Authority under the CBA, it would appear that even where the signal is encrypted (provided decoding devices are generally available to the public) these transmissions are treated as for general reception so that the broadcasting authority concerned is excused from the requirement to obtain a telecommunication licence for such activities.[5]

C. VALUE ADDED

"Value Added Services" is an expression used relatively loosely to define that category of services involving the control and processing of messages, over and above their basic transmission. One regulatory definition is contained in Condition 4.8 of the VADS Class Licence, but aside from its application to the use of international private leased circuits this definition has no real significance. It is more important to ascertain whether the provision of any particular service of this nature by means of a telecommunication system is covered by the BSGL or the VADS Class Licence. If the BSGL or the VADS licence conditions or criteria cannot be met, an individual licence for the running of the system will be necessary.

The licensing of value added and data services is discussed in detail above.[6]

D. MOBILE

1. Mobile telecommunication services involve the use of wireless telegraphy for the transmission to and reception from users (typically of mobile telephones or paging devices) of messages. Accordingly, as discussed above,[7] a licence under the Wireless Telegraphy Act 1949 is required for such services but in addition a licence under the Act is obligatory. Licences have been granted for only two cellular radio telephone operators, Telecom Securicor Cellular Radio Limited (Cellnet) and Racal Vodafone Limited and for a limited number of mobile radio telephone organisations.

2. Cellnet and Racal Vodafone's licence service obligations are similar to Mercury Communications in that each is obliged to develop a national system but over a period which will allow for rational development. Accordingly, the requirement of it is that by December 31, 1989 service should be available to its customers over an area where 90 per cent. of the United Kingdom population live. Cellnet and Racal Vodafone is each also obliged to take all reasonable steps to ensure the provision to its customers of international services, although this will have to be done via BT or Mercury as the only licensed providers of international services. Racal Vodafone has an interconnection agreement with BT and Mercury respectively.

Cellnet and Racal Vodafone are prohibited from engaging in the business of producing or supplying telecommunication apparatus or of providing value added services.

Cellnet and Racal Vodafone's licence sets up a special regime for distribution of their services. Condition 12 of its licence provides that they must provide their services through people engaged in the business of providing retail telecommunication services. This presumably was intended to encourage the rapid development and deployment of the services and anyway is a technique which is used voluntarily as a marketing tool in other areas of the provision of telecommunications services, for example radio paging.

3. As with mobile radio telephone services, licences have been granted for wide area radiopaging services to a limited number of applicants. Typically these services involve the ability to communicate a message to the user of a paging device, by the sender of the message communicating first with a computer over the public switched network. A message is then passed from the premises where the computer is located over the fixed links of a PTO, to reach a transmitting station which, by wireless telegraphy, communicates to the paging device. Standard devices emit a sound or flashing light to warn the carrier of the device that a message has been sent to him and more sophisticated devices can provide an abbreviated message on a screen.

By virtue of these arrangements for conveyance of

messages, operators of radiopaging systems will inevitably require an interconnection agreement with one or both of the PTOs.

As with mobile radio licences, typically the licence granted to a radiopaging operator contains obligations with regard to expansion of the coverage of the service nationally. The licences also contain obligations for the protection of consumers, such as the requirement to publish a Code of Practice giving guidance to customers with regard to disputes and complaints and provision for arbitration of disputes with customers. There is, however, normally no requirement on radiopaging licensees to publish charges and other terms and conditions of service.

E. CABLE TELEVISION

1. Systems for the communication of video messages and associated signals again require a licence under the Act. Where these messages represent a cable programme service[8] a licence will also be required from the Cable Authority, under Section 4 of the CBA. However, it is the practice that one application to the Cable Authority suffices for both licences.

2. Specifically for cable television, franchised areas have been and continue to be allocated in various parts of the United Kingdom and licences granted to operators of cable television (broadband) systems in these areas. These licences represent another limited exception to the BT and Mercury monopoly over the provision of fixed links in that television operators are licensed to install their own cables and run them from the "head end" to users' homes. Clearly this would involve the digging up of streets and other activities similar to those of BT and Mercury and accordingly it has been necessary to designate cable television operators as public telecommunication operators to whom the provisions of the Telecommunications Code[9] will apply. The licences granted to the long-line PTOs, BT and Mercury, specifically preclude their providing telecommunication services which are cable programme services.

3. According to the Cable Authority, franchises will be granted to those prospective operators who appear best placed to exploit the potential of cable and to deliver the benefits it offers. The matters to be taken into account by the Cable Authority are set out in section 7 of the CBA, particularly with respect to programming, and the CBA will also base its judgment on the type of technology to be used by the applicant, its financial resources and its management and marketing skills.

The procedure for granting a cable franchise is set out in sections 5 and 6 of the CBA and essentially is as follows:

(a) The Cable Authority publishes a notice stating that it proposes to grant a licence for the provision of a "prescribed diffusion service"[10] in the particular area and inviting applications for the licence;
(b) The Cable Authority then publishes details of the applications it has received;
(c) The Cable Authority must canvass local opinion as to the services to be provided in the area and will also consult the DTI and OFTEL with regard to the telecommunications licensing aspects, as well as consult every local authority whose area is affected by the applications.

This process may take up to about four months, in order for the Cable Authority to announce its decision. Even then the decision will be conditional upon the grant of a licence under the Act. The Secretary of State will also go through the procedure for application of the Telecommunications Code to the licensee and its designation as a PTO.[11]

4. The licence from the Cable Authority will be for a period of 15 years from the start of services on the system, with provision for renewal for eight years at a time, but only after the Cable Authority has re-advertised the franchise and considered all the applications received. The telecommunications licence granted to the operator will be for either 15 years or 23 years depending on the nature of the system. Some telecommunication licences for cable television will be of only 15 years duration at the outset, but with the possibility of extension if the licensee commits himself to upgrading his system within a specified period. The technical requirements meriting the grant of the longer licence are normally set out in Condition 13 of the licence and would include:

(i) the customer's ability to make independent selection of two or more channels;
(ii) a system capability of providing data services to all who request them and are prepared to pay for them;
(iii) the capability of providing a return video channel on request to no more than one subscriber simultaneously and
(iv) a commitment eventually to permit persons, who want to make use of such a facility and are prepared to pay for it, to send and receive data signals at a rate of 64 kilobits per second and to ensure that such a customer can have access from the cable system to the public switched network.

In addition to a £2,000 application fee further fees would normally be payable to the Cable Authority in stages depending on "homes passed" by the cable system.

5. There are certain restrictions on the persons who may hold, and therefore apply for, licences under the CBA. These provisions for example exclude from eligibility any company incorporated outside the EEC, an individual resident outside the EEC, local authorities, political or religious bodies, local independent television and independent radio contractors and local newspaper proprietors. There are also provisions to allow revocation of the licence if there is a change in the make-up or control of the franchise holder after the licence has been granted.

United Kingdom incorporated companies with overseas, *i.e.* non-EEC, shareholders are not precluded from becoming cable television franchisees and holders of telecommunications licences for this purpose provided their stake in the company does not exceed 50 per cent.; the government has indicated that where ownership is fragmented it would normally expect non-EEC community participation to be no more than 30 per cent. Further guidance on all these requirements and on, for example, the content requirements of programme services, can be obtained from the Cable Authority itself.

F. SPECIALISED SATELLITE SERVICES

Satellite uplinks and downlinks fall in principle within the duopoly of BT and Mercury, destined to last until at least November 1990. However, the Kenneth Baker statement of November 1983[12] included a passage that "the Government will keep under consideration ways of introducing new specialised services by satellite"; during 1987 and 1988 the Government was considering the implications of this statement in the light of representations it has received regarding liberalisation of satellite services. At the time of writing the Government has issued a statement[13] indicating its intention to license up to an additional six operators to provide one–way point to multi––point satellite–based third party services in the United Kingdom only. These services could cover voice, data and video/entertainment services (subject to provision of any necessary CBA licences).

The same statement announced the Government's intention to allow the use of reception dishes for satellite signals of all kinds, for which a class licence under the Act would be issued. A suitable exemption under the WTA would also be granted.

G. INFORMATION AND ENTERTAINMENT SERVICES

1. This category is intended to describe the burgeoning industry in the provision of all kinds of non-telecommunication service, from the Speaking Clock to entertainment (for example pop music) to financial information (for example BT's Citycall).

As mentioned previously, there are few rules governing such information services and normally the system being run for the purpose would be covered by the BSGL.

These services fall within the scope of cable programme services under the CBA and therefore to avoid the anomaly of a licence being required from the Cable Authority an order was issued in 1984, The Cable Programme Services (Exceptions) Order 1984, under provisions of the Act which have now been repealed and re-enacted in the CBA, in respect of the various services referred to in the Schedule to that Order. These include cable programme services "consisting only of the sending in sound only of recorded matter of a repetitive nature."

2. In 1985 BT introduced its premium rate service (now known as "Callstream") which involved its customers paying a premium rate for the service of having their messages conveyed to information service providers. Under BT's typical contractual arrangements with service providers BT pays them a percentage of the gross revenue which it receives from the calling customers. These percentages currently vary according to whether national or London local numbers are involved.

3. It was through various service providers' dealings with BT that it was found convenient that they should present a uniform and collective approach to BT on certain matters which were common to them all. This resulted in the creation of Association of Telephone Information and Entertainment Providers Limited ("ATIEP") in July 1986. ATIEP's central aim is to represent the interests of service providers and to maintain standards within the industry. ATIEP is open for membership to any service provider, provided it is of sufficent size (the current minimum entry requirement being 32 lines). Until now its principal activities have been in negotiating prices, percentages and the form of standard contract with BT on behalf of its members. It also drew up, shortly after its formation in 1986, in consultation with BT and OFTEL, a Code of Practice for the industry and is active in negotiating with music collecting societies blanket licence terms on which service providers can broadcast music.

ATIEP's Code of Practice obliges service providers to comply with the British Code of Advertising Standards and also requires conformity with the IBA's Statement on "family viewing policy" for material broadcast on television prior to 9 p.m. That policy includes important safeguards covering messages specifically targetted at children, which cannot be of a sexual or violent nature and must not "encourage insulting behaviour, rowdiness, impertinence or bad manners."

Currently observance of the Code of Practice is supervised by the Independent Committee for the Supervision of Telephone Information Services headed by barrister Louis Blom-Cooper.

The sending of messages as part of the provision of such services is governed by the general rules as to the use of telecommunication systems (for example section 43 of the Act in relation to indecent, obscene or menacing messages etc). Other rules, for example as to copyright, will require consideration: the playing of records over telephone lines amounts to their public performance and accordingly the licence of the Performing Rights Society and of Phonographic Performance Limited would be required. If such a service involves making a recording of music the licence of the Mechanical Copyright Protection Society will also be required.

Note on Radio Frequency Allocation

In April 1987 the Government published a report "Deregulation of the Radio Spectrum in the U.K.,"[13] forecasting demands on the various frequencies and frequency bands into the 1990s. The report recommends the introduction of competitive tendering for spectrum allocation. A bill on this whole subject may be introduced in Parliament in 1988.

Notes

[1] See Chap. 1 at p. 11 and Appendix A.
[2] Chap. 6, at p. 96.

[3] See Chaps. 5 and 6 at pp. 80 and 96 above.
[4] See Government announcement on Liberalisation of Specialised Satellite Services February 17, 1988; also F. at p. 121.
[5] See s.6(1) of the Act.
[6] See Chap. 6 at p. 96 above.
[7] See Chap. 2 at p. 28.
[8] S.2(1) C.B.A.
[9] See Chap. 10 at p. 151 below.
[10] s.2(3) C.B.A.
[11] See Chap. 3 at p. 42.
[12] See n. 1 above.
[13] February 17, 1988; see n. 4.

Chapter 8

APPARATUS

Apparatus is a matter for regulation in order that reasonable safety standards should be met, that the intergrity of public networks should be maintained and that in particular, using the general criteria set out in Condition 14 of BT's Licence and reflected in other PTO licences, such networks should suffer no "material impairment" as a result of their connection to any apparatus.

All licences issued under the Act for the running of telecommunication systems have a term (usually set out in Annex A) requiring that:

> "If the system is connected (whether directly or by means of a telecommunication system other than a Specified Public Telecommunication System) to a Specified Public Telecommunication System, the apparatus comprised in the system is approved under section 22 of the Act for connection to:
>
> (i) the system; and
> (ii) in the case of apparatus also directly connected to a Specified Public Telecommunication System, that System."

There is a further provision common to all licences which authorises the connection of the licensed system to other apparatus provided that apparatus is itself comprised in the licensed system or another licensed system and is therefore approved under the Act.

Apparatus comprised within a system which is exempt from this licence requirement does not require to be approved under the Act.[1]

Otherwise, it must be remembered that approval of apparatus for telecommunication systems is vital in that a person running a licensed system will be operating outside that licence and committing an offence under section 5 of the Act if the licensee incorporates in the system or connects to it any unapproved apparatus.

Section 22 of the Act provides that apparatus may be approved either by the Secretary of State or (with the consent of, or in accordance with a general authorisation given by, the Secretary of State) by the Director General. This general authorisation to the Director General has now been given.

The Director General is required to keep a register of approvals given and designations made under section 22 and in practice he also regularly publishes details of recent approvals (see OFTEL's "UPDATE").

In granting approval to apparatus, the Director General's discretion under his general authorisation is more limited than that of the Secretary of State (*e.g.* by virtue of Condition 14.2(b) of BT's Licence) in that the Director General must be satisfied, before approving any apparatus or designating any standard to which apparatus must conform, that it would not be liable to cause death or personal injury or damage to property, nor materially impair the quality of any telecommunication service provided by means of a system to which the apparatus is to be connected.

Accordingly, if the Director General cannot be so satisfied, approval can only be obtained from the Secretary of State himself.

1. Telecommunication Approvals Manual

A private company, Communications Educational Services has, in association with OFTEL, published a "United Kingdom Telecommunication Approvals Manual" ("the Manual") written by another private organisation, Interconnect Communications which is a semi-official (in that it is endorsed by OFTEL) bible of rules, practices and procedures for apparatus approval. It is thus an essential reference for anyone concerned with the detail of such approvals.

2. Categories of Approval

According to the Manual apparatus approval can only be applied for by concerns resident or having their place of business in this country. It seems likely that the main or only reason for this is the requirement imposed upon suppliers of approved apparatus to ensure that the terms of the approval are followed and that for example the apparatus is not modified so as to negate the validity of the approval. It is questionable whether this requirement would stand up to legal scrutiny, particularly in relation to European suppliers who might well be able to allege that it amounted to a restriction on imports in contravention of Article 30 of the Rome Treaty. In practice, however, it would appear that OFTEL is prepared to issue approvals in the name of foreign manufacturers provided they can give a "care of" address in the United Kingdom, typically that of their United Kingdom distributor or sales agent.

In regulatory parlance, there are essentially three different categories of approval:

(a) General approval: as the description implies, this covers all apparatus conforming to the description set out in the approval; accordingly no evaluation or testing process would be required. One example of a general approval is that applicable to certain indirectly connected non-speech apparatus (see OFTEL Update SA/6), which applies to a wide range of data terminating apparatus including personal computers. It allows them to be indirectly connected, without further technical evaluation, to public switched networks and private circuits, provided the connection is via an item of type-approved apparatus incorporating an approved barrier device.

There is now a General Approval allowing attachment to the Mercury System of any apparatus approved for connection to the BT Systems on or before September 8, 1987.[2] After that date appropriate approvals automatically cover Mercury's System.

(b) Type approval: this certainly requires the evaluation process to be followed because once given it applies to every example of the particular type of apparatus. In such a case the evaluation body (see below) will require to be satisfied not only as to the performance of the samples submitted to it but also on the manufacturing and quality control procedures to be followed by the producers of the apparatus.

(c) One-off approval: a one-off approval would typically be applied to apparatus which remains to be fully developed or is only to be produced in small quantities. As a consequence the life of one-off approvals is normally fairly brief.

In June 1987 OFTEL announced that BABT had introduced a pilot scheme for approval of a wide range of telecommunications attachments. The Scheme applies initially to telephony apparatus and certain auxiliary apparatus only and will later cover modems. Under its provisions (explained in OFTEL's Update SA/8) the evaluation procedure is divided into a defined sequence of events essentially involving testing by one of three test laboratories leading ultimately to a laboratory report to BABT on which BABT conveys its recommendation for approval to the Director General.

3. Evaluation

Having identified the type of approval required, the applicant would have to submit his request for approval to the relevant evaluation authority which in most instances would be BABT. Apparently if BABT consider that the apparatus in question falls outside the ambit of its standard procedures, the recommendation will normally be made that the applicant seeks to have the evaluation performed by a PTO which currently would almost certainly only be BT as the other PTOs, such as

Mercury and Hull, have yet to specify their requirements.

Evaluation by an authority like BABT is carried out by reference to Technical Information Sheets ("TISs") which identify the requirements, including any Special Investigation Test Schedules ("SITS") where such exist. If there is no SIT which can be used, BABT may be able to prepare one for the specific apparatus, but this would have to be brought before the BSI Technical Committee for their endorsement.

Needless to say BABT's evaluation is not done for free and its fees are based on the actual testing charges and expenses, with a mark-up to cover BABT's own involvement. Apparently BABT also offers consultancy services at fixed hourly rates.

Apart from BABT, BT is also involved in evaluation of apparatus, though to a diminishing extent as gradually this is all being taken over by BABT. Apparently BT's evaluation work is largely in relation to the more unusual types of apparatus, presumably particularly those which have potentially significant effects on the public networks.

In order to initiate evaluation of apparatus by BT, at the time of writing it is BT's policy to require the applicant to enter into an agreement containing an undertaking not to make any changes to that apparatus and an additional undertaking to enter into a "suppliers agreement" amongst other things indemnifying BT against loss caused by the apparatus. In fact this agreement in its current form goes some way further than this without any explanation as to why this should be so. The author must therefore question why this practice has developed for two fairly evident reasons. The first is that an unapproved change to the apparatus would render invalid the original approval granted by the Director General. The second is that surely the main purpose of the approval process is to ensure that the approved apparatus will not damage a public network such as that of BT; if it did either BT (Teleprove) has itself to blame or the approval holder has been negligent in which case BT would have a right of action in any event.

Standards are gradually being developed for apparatus by the British Standards Institute. If a standard has not been developed, the approval authority (at present the DTI or the Director General) must be satisfied as to the safety of the apparatus and the fact that it should not materially impair service, as explained above. Approval authorities will also be concerned to satisfy themselves that the apparatus complies with the United Kingdom's obligations under the European Communities Council Directive No. 73/23/EEC on the harmonisation of the laws of the Member States relating to electrical equipment designed for use within certain voltage limits.

4. Marking and Advertising

Section 28 of the Act gives the Secretary of State authority to make orders requiring telecommunication apparatus to be marked with certain information or instructions relating to the apparatus or its connection or use. He has done this under the Telecommunication Apparatus (Marking and Labelling) Order.[3]

Under this Order which applies only to apparatus supplied through retail outlets, approved apparatus must carry a solid green circle together with details of the approval number if given and a legend referring to the fact that the apparatus is approved for connection to telecommunication systems specified in the instructions for use of the apparatus.

The Order goes on to provide that unapproved apparatus must carry a solid red triangle referring to the fact that it is prohibited from connection to public networks and again with a legend to this effect.

Section 29 of the Act authorises the Secretary of State to make orders as to the content of advertisements for telecommunication apparatus. The Telecommunication Apparatus (Advertisements) Order[4] provides that advertisements for telecommunication apparatus subject to the order must, with limited exceptions, display the fact of their approval. The detailed requirements vary according to the advertising medium used.

5. Network Code of Practice

Approved apparatus goes to make up the telecommunication systems which are run in order to provide telecommunication service or so as to be connected with other systems providing such services. The Director General is concerned that all systems connecting with public switched networks should give certain minimum standards of quality and these standards in turn depend upon the links, switches and terminal apparatus used for various routings in the relevant system. This concern with standards has been translated into licence obligations, for example in the terms of Condition 4(b) of the BSGL.

The technical requirements to which Condition 4 refer are to be gradually developed by the Director General in relation to connections with various public networks. The first attempt at formulation of standards for this purpose is in the "Provisional Code of Practice for the Design of Private Telecommunication Branch Networks" published by OFTEL in December 1986 (the "Network Code of Practice" or "NCOP").

According to OFTEL, the code has been issued as a provisional document and will not have binding effect until work on it has been completed and OFTEL has had the opportunity to assess its operation in practice. In the meantime, the standards set in the NCOP are being adopted by OFTEL as an indicator of the standards required to avoid material impairment of public networks. This is directly relevant to Condition 14.3 of BT's Licence which allows BT to refuse to connect or to disconnect apparatus which in BT's opinion is liable *inter alia* to cause material impairment to the quality of its telecommunication service, provided "the Director has not expressed a contrary opinion."

There is in fact no reason why BT should adopt the standards in the NCOP as its own indicator of material impairment although in practice it must be hoped that

BT's and OFTEL's views on this will be consistent with each other.

The NCOP lays down performance standards for branch networks in order that satisfactory call quality can be achieved. Recognising that this depends on the degradation of performance for each element of the network, including links, switches and terminal apparatus, it requires that the sum of the degradations encountered on any call meets the NCOP standards.

The NCOP applies to connections to public switched systems providing telephony services. (Apparently supplements to the NCOP will be issued in the future dealing with access to other public switched networks including telex.)

Although primarily concerned with design criteria applying to public switched network traffic, where the requirements of the NCOP are mandatory, it also makes recommendations, which are not mandatory, for calls wholly within a private network.

The NCOP identifies various design characteristics of a private branch network which are significant in the context of call quality namely:

Safety
Address structure
Call control
Call progress indication
Call path transmission quality
Traffic handling capacity.

In addition, in sections 2 and 3 of Annexe 3 of the NCOP mandatory requirements are included for:

(i) Overall loss
(ii) Absolute delay
(iii) Echo and stability
(iv) Quantising distortion
(v) Signal power level and out of band noise
(vi) Jitter and wander
(vii) Synchronisation
(viii) Coding standards.

An all-digital network with an appropriate signalling system, PSTN-type tones and CCITT recommended grade of service levels should, not surprisingly, apparently satisfy NCOP requirements. If any of these elements are not satisfied to any degree the usability of the network becomes more constrained.

The procedure for implementation of the NCOP will require refinement and is already controversial in the way in which it involves public operators in the compliance process.

The person responsible for ensuring compliance with the NCOP appears to be the licensee running the private branch network, who must provide a Certificate of Compliance (a specimen of which is annexed to the NCOP) to the operators of those public switched telecommunication systems to which the branch network is connected.

According to section 5.1 of the NCOP this certificate is, in the case of a company, to be signed "by an officer personally responsible" for the discharge of its legal obligations. This is somewhat vague and it must be hoped will be specified in more explicit terms in the final version of the NCOP.

The person giving the certificate of compliance must also prepare and retain a network plan and in the case of reasonable doubt as to compliance with the NCOP it is provided that the relevant PTO may "require access to this plan." There is in fact no legal right for the PTO to demand access to this plan although in practice a PTO may be able to obtain the plan on the basis of the licensee otherwise risking a report to OFTEL that the licensee may be in breach of its licence. Concern has been expressed in certain quarters that the PTO's part in this compliance process can potentially give a PTO access to service providers' leased circuit network designs, which is commercially sensitive if they are competing with the PTO.

On October 14, 1987 a related OFTEL proposal, a Temporary Scheme for site specific approval, to allow the connection of private networks to the PSTN, was published

(UPDATE SA12). This is designed to allow BSGL licensees to convey within their own private networks messages which have come from a public network. Under the Scheme (to last until October 1989) suppliers of CRAs are to declare their impairment values so that the private network operator may determine that his network complies with the NCOP. If the network is compliant OFTEL would then issue a site-specific approval for all apparatus in the operator's network.

6. Connection of Apparatus

The arrangements for physical connection of apparatus to telecommunication systems are essentially to be found in the "General Conditions Applicable to Licences Granted under section 7 of the Telecommunications Act 1984" issued by the Department of Trade & Industry on February, 25 1987. These General Conditions[5] have been adopted for the BSGL and the VADS Licence and are being adopted for other individual and class licences issued under the Act.

The two conditions which are likely to prove of most interest in that they deal with connections to public telecommunication systems are Conditions 5 and 11.

Condition 5 provides that a licensed system may only be connected to a public system by means of Network Termination and Testing Apparatus which is:

> "(a) comprised in the Specified Public Telecommunication System
> (b) installed on premises occupied by the Licensee and
> (c) supplied, installed, Brought into Service and maintained by or on behalf of the operator of the Specified Public Telecommunication System to which the connection is or is to be made."

This, therefore, deals with the status of the network termination and testing apparatus. So far as the physical act of connection is concerned, Condition 5.2 provides that unless the connection can be made by use of a plug and socket, the licensee cannot himself break, or authorise anybody else to break, any tangible connection between his system and that of the public operator, unless the authorisation of that public operator is obtained or that public operator actually carries out the work.

In practice, there may be mere millimetres between the actual interface of the branch system with a public network and some other component at which point a connection may be temporarily broken to accommodate certain apparatus. Accordingly there is scope for dispute in interpretation of Condition 5.2 in that it is not entirely clear whether in such circumstances the connection which is broken would be deemed to be remote from the connection with the public network, in other words whether for these purposes "connection" means direct connection at the interface with that network.

Condition 11 relates to the connection of telecommunication apparatus, which is not itself directly connected to a public system, to an item of Call Routing Apparatus. Call Routing Apparatus is defined in the General Conditions (see above) as apparatus *"installed and connected* so as to be *capable"* of switching two-way live speech telephone calls between two or more internal extension and two or more public system circuits. Thus if apparatus has that capability but is not installed and connected in order to realise it, the apparatus shall not be classed as Call Routing Apparatus.

Condition 11.1 provides that such apparatus shall only be brought into service by either the Designated Maintainer of the Call Routing Apparatus or "any other person" where the Designated Maintainer has so agreed or has himself failed to bring the apparatus into service within 14 days of notice from the licensee. Designated Maintainers are either public operators (for example BT, Mercury or Hull) or approved contractors (under section 20 of the Act, with whom the relevant licensee has a contract for the provision of maintenance services). Maintenance is discussed below.

The whole aspect of the connection of apparatus to public systems is a sensitive area and as indicated above, problems are likely to arise from time to time over whether or not a connection between a relevant system and a public system is being broken and as to whether or not a licensee can exercise its right under Condition 11 to use some other person (who appears not to require any qualification whatsoever) to bring apparatus into service where the Designated Maintainer has not complied with the 14–day notice.

7. Equipment Procurement: EEC requirements

Procurement of telecommunications apparatus is subject to a Recommendation from the Council of Europe[6] which requires network operators throughout the Community to operate open tendering procedures for all new terminal apparatus where common type approval specifications exist, and for 10 per cent. by value of total annual orders for switching and transmission apparatus and conventional terminal apparatus. In the United Kingdom this Recommendation applies only to BT, Mercury and Hull as PTOs.

8. EEC Proposals on Common Standards and Specifications for Telecommunications Equipment etc.

For some time the European Commission has been eager to establish common standards and specifications for telecommunications equipment, in order that the Community's market in such equipment should become fully liberalised. On July 24, 1986 the Commission issued a directive[7] on the initial stages of mutual recognition of type approvals.

During the summer of 1988 the Commission is due to issue a Directive[8] regarding liberalisation of the terminal equipment market.

Notes

[1] See Chap. 3 above, at p. 35.
[2] NS/G/0001/G/100001.
[3] S.I. 1985, No. 17; amended by S.I. 1981, No. 1031.
[4] S.I. 1985, No. 719, amended by S.I. 1985, No. 1030.
[5] Discussed in Chap. 5 above, at pp. 89 *et seq*.
[6] 84/550/EEC; November 12, 1984.
[7] 86/361/EEC.

CHAPTER 9

COMPETITION LAW

This area of telecommunications regulation is a hybrid of specially formulated legislation and quasi-subordinate legislation and existing fair trading/competition laws. It is characterised, as is all United Kingdom competition law, by a lack of direct enforceability and the requirement, under our domestic rules, for administrative intervention.

1. Regulation under the Telecommunications Act 1984

As has been seen,[1] the licences of public telecommunication operators and of Major Service Providers under the VADS Licence contain conditions imposing reasonably extensive, but selective, prohibitions on certain aspects of anti-competitive behaviour. These notably include undue discrimination, linked sales (otherwise known as "tie-ins"), exclusive dealing arrangements and restrictive use of intellectual property rights. Enforcement of these licence conditions is primarily a matter for the Director General under sections 16 to 19 of the Act. Study of these sections will show that the procedure for bringing a recalcitrant operator to heel is somewhat long-winded and may in certain circumstances prove to be insufficiently speedy to provide an effective remedy. To summarise this procedure, where the Director General has investigated a possible breach of a licence condition by an operator (*i.e.* person licensed to run a system) and is satisfied that the operator is in breach the Director General can do one of two things. If he fears that the breach will continue and, in particular, may cause loss or damage to any person, he may make a provisional order to secure the operator's compliance with the particular licence condition. Clearly a provisional order is intended to be temporary and by virtue of section 16(7) of the Act it ceases to have effect at the end of a maximum period of three months unless it has been previously confirmed.

Otherwise, and instead of a provisional order, the Director General may proceed directly to the making of a final order. The procedure for such an order, which also applies to confirmation of a provisional order, is that the Director General should first give notice to the operator concerned that he proposes to make or confirm the order and setting out its effect. Such notice should state the relevant conditions of the licence which the operator is alleged to have contravened and give a time, at least 28 days from the date of publication of the notice, within which the operator may make representations on, or objections to the proposed order.

If notwithstanding these representations or objections the order is confirmed or made, the operator may under section 18(1) of the Act apply to the Court for determination of the question whether it was within the powers of the Director General to make the order or whether he complied with the procedural requirements mentioned above. If the Court finds that the order was not within the Director General's powers or that he did not follow the correct procedure, it may quash the order or any relevant provision which it contains. Section 18 goes on to state that except as it provides, "the validity of a final or provisional order shall not be questioned by any legal proceedings whatever."

All this may appear relatively unremarkable. However, an order once made can have severe implications: the obligation to comply with the order is a duty owed to any person who may be affected by its contravention. Any such person is able to take action not only against a defaulting operator breaching an order but also against any person guilty of an act which induces the operator's breach or interferes with its performance of the order. In any such proceedings pursuant to section 18 it will be a defence to prove that the defendant "took all reasonable steps and exercised all due diligence to avoid contravening the order."[2]

These particular provisions giving rise to liability for inducement of a licence breach caused particular consternation to BT's employees' trade unions during the passage of the Telecommunications Bill. Prior to its enactment, the prevention of transmission of telecommunication messages had been a criminal offence; in the event this criminal liability was removed and substituted by the civil liability imposed by section 18(6). This sub-section was said to be drafted with particular care so that a person inducing a breach is only liable if his act is done wholly or partly for the purpose of achieving the result of causing the breach. Accordingly, it will be for the aggrieved person seeking a remedy to prove that that was indeed the purpose with which the act was done. The Director General is to keep a register not only of licences granted but also of final and provisional orders made under the relevant sections of the Act.

Breach of a licence condition should be distinguished from failure to adhere to the provisions of the licence grant itself, for example the permitted services in Schedule 3. Such a failure would put the operator's activities beyond the protection of the licence so that, in therefore running an unlicensed system the operator becomes guilty of a criminal offence.

2. Inherited Functions—Fair Trading Act 1973 and Competition Act 1980

Whatever his rights and powers in any particular case, the Director General has a general duty under section 49 of the Act to investigate complaints in relation both to the supply of telecommunication services and of telecommunication apparatus.

Section 50 of the Act specifically transfers to the Director General the functions which were conferred upon the Director General of Fair Trading under the 1973 Act in relation to the investigation and control of monopoly situations in the telecommunications industry. Similarly, the functions of the Director General of Fair Trading under sections 2–10 ("Control of Anti-Competitive Practices") of the 1980 Act have been transferred to the

Director General in respect of telecommunication services and apparatus.

These functions are in both cases, under the 1973 and the 1980 Act, to be exercised "concurrently" with the Director General of Fair Trading and if either of the two Directors wishes to take any action under either piece of legislation he is first to consult with the other Director.

Neither Director is to exercise such functions if the same functions have been exercised by the other Director in relation to the same matter.[3]

Under section 50 of the Act, the Director General must exercise the Director General of Fair Trading's functions as to courses of conduct detrimental to consumers of telecommunication services or apparatus, if and when requested to do so by that Director General.

Licence Modification: Under section 95 of the Act, where, following a monopoly, merger or competition reference, the Secretary of State exercises any of his powers under Parts I and II of Schedule 8 to the Fair Trading Act, he may also provide for the revocation or modification of a licence granted under section 7 of the Act. This would enable him, for example, to amend the telecommunication licence of a PTO which had been the subject of a particular reference of this type and this power would appear to extend to all provisions of such a licence, not just its conditions. By contrast, licence modifications would of course be more usually effected pursuant to sections 12 to 15 of the Act,[4] when such modifications are limited to the licence conditions.

3. Monopoly and Merger References under the 1973 Act

a. Monopoly references

A monopoly reference can be made by the Director General, on the basis and in accordance with the procedure set out in sections 47 to 56 of the 1973 Act, but only in respect of the supply of telecommunication apparatus, not in relation to the running of a telecommunication system.[5] A reference in respect of tele-

telecommunication services would therefore have to be made by the Secretary of State.

The Director General may make a monopoly reference at any suitable time, but if he chooses to do so it would normally be because he was concerned about a particular practice or course of conduct which was "uncompetitive" within the context of section 48 of the 1973 Act.

There are two possible types of monopoly reference, one "limited to the facts" and the other "not limited to the facts." There have been very few references limited to the facts. As regards those not so limited the MMC may only investigate and report on such a reference with a view to determining whether or not a monopoly situation exists and whether any act or omission complained of in the reference operates, or may be expected to operate, against the public interest. Such matters would have to relate to prices or recommendations or suggestions as to prices, refusal to supply and preference "whether by way of discrimination in respect of prices or in respect of priority of supply or otherwise"; this latter concept is broader than the undue discrimination provisions in BT's Licence.

Under section 55(1) of the 1973 Act the MMC must normally report on a monopoly reference within the period specified in the reference. (The period can be extended more than once and there is no overall limit). If the report concludes that any matters do operate against the public interest the Secretary of State may by order made by statutory instrument exercise any of the powers specified in Parts I and II of Schedule 8 to the 1973 Act. These powers are wide-ranging but essentially enable the Secretary to proscribe the offending activities in relation to prices, refusals to supply, preference and discrimination. Part II powers are only exercisable with the approval of Parliament.

In passing, it should be noted that under section 78 of the 1973 Act the Secretary of State may at any time require the MMC to report on the general effect, on the public interest, of practices which are "commonly adopted" in order to preserve monopoly situations or on practices which appear to be "uncompetitive practices." He may also require the MMC to report on appropriate remedial action required.

a. Merger references

Merger references remain the preserve of the 1973 Act[6] and the Secretary of State. He may refer to the MMC a "merger situation" qualifying for investigation, being a situation where two or more enterprises, one of which is carried on in the United Kingdom have "ceased to be distinct" and either:

(a) as a result, at least a quarter of the relevant goods or services are supplied by or to, one and the same person or such a previously existing market share has been increased; or
(b) the value of assets taken over exceeds (currently) 30 million.

The most recent example of such a reference in the telecommunications field was the then proposed merger of BT and Mitel Corporation, on which the MMC reported in January 1986.[7]

4. Competition References—1980 Act

The transferred functions of the Director General under sections 2 to 10 of the 1980 Act empower him to investigate anti-competitive practices, to report on the results of such investigation, to accept undertakings in relation to any anti-competitive practices the subject of such report, and to refer any such practice to the MMC. Following an MMC report,[8] on the request of the Secretary of State he may seek undertakings from the person concerned to take or refrain from taking action with a view to remedying any effects of such practice found by the MMC to operate against the public interest. For these purposes an anti-competitive practice is, for once, defined in terms similar to those used in Article 85 of the Rome Treaty, although the ambit appears to be narrower in that the practice must amount to a "course of conduct." In context, under section 2 of the 1980 Act, and substituting the Act's amendments, a person is said to engage in an anti-competitive practice:

> "If in the course of business that person pursues a course of conduct which, of itself or when taken together with a course of conduct pursued by persons associated with him, has or is intended to have or is likely to have the effect of restricting, distorting or preventing competition in connection with the production, supply or acquisition of *telecommunication apparatus* in the United Kingdom or any part of it or the supply or securing of *telecommunication* services in the United Kingdom or any part of it."

Such anti-competitive practices have their own near equivalent to the Article 85 (Rome Treaty) minor agreements exemption. Under the Anti-Competitive Practices (Exclusions) Order 1980,[9] a course of conduct is excluded from constituting an anti-competitive practice if it is:

(a) a course of conduct described in Schedule 1 to the Order; this includes contracts for exports (where of course any such conditions affecting EEC trade could still merit review in relation to Article 85) and
(b) the course of conduct of a person whose (or whose group's) annual turnover in the United Kingdom is less than 5 million and who (individually or whose group) enjoys less than one-quarter of a relevant market (*i.e.* in respect of the particular goods or services involved).

Although the language of section 2 of the 1980 Act, in describing an anti-competitive practice, is similar to its Euro-equivalents, the similarity becomes less apparent on examination and comparison of the procedure adopted by the 1980 Act to deal with such practices. Whereas, for example, the European Commission can investigate and issue legally binding pronouncements as to the validity of agreements and practices within the bounds of Article

85, in some cases giving rise to directly enforceable claims by persons affected, the Director General can undertake investigation but cannot take any executive action himself. Once the matter has been referred by him to the MMC it is out of his hands unless and until he is to seek undertakings (if any) arising out of the MMC report.

The MMC's terms of reference and its ability to make suitable recommendations to bring an end to an anti-competitive practice are heavily circumscribed. In particular, on both a monopoly reference under the 1973 Act and a competition reference under the 1980 Act, the MMC's objective is to discover whether or not there are effects adverse to the public interest. If the public interest is not adversely affected, the MMC has no locus to make any other determination on the matter. This emphasis purely on the public interest criterion appears to be one peculiar to United Kingdom anti-trust and it would have been preferable in terms of potential harmonisation of European Competition Laws and certainly preferable for those seeking to prevent anti-competitive practices or who might be prejudiced by them, for the test simply to be whether or not the particular situation gave rise to a significant or appreciable restriction on competition or represented the abuse of a dominant position.

Thus it may come as a surprise, at least to foreign lawyers, that under the 1980 Act an anti-competitive practice is not *per se* illegal and does not invite legal sanction until the whole procedure of Director General investigation, MMC report and Secretary of State order has been completed.

5. European law: The Rome Treaty

This is not the place to go into a detailed examination of the texts of Articles 85 and 86, which are well rehearsed in other publications on the specific subject of European competition law. For present purposes it suffices to note that many of the Fair Trading provisions contained in PTO licences and the VADS licence are directed against practices which, where there is an appreciable effect on inter-Member State trade, would also be liable to fall foul of Article 85 or, where a dominant position is being exploited, of Article 86.

These Articles have already become a potent force both in the hands of the European Commission and in their practical influence on contractual relations between parties of unequal status or bargaining power, or who might otherwise be subjected to competitive restraints.

Article 85: This Article prohibits agreements and concerted practices which may affect inter-member state trade and which have as their object or effect the prevention, restriction or distortion of competition within the Common Market.

Article 86: This Article prohibits an abuse by an undertaking of a dominant position within the Common Market. It gives specific examples of such abuse, such as the imposition of unfair prices or other unfair trading conditions, the limitation of markets, the application of dissimilar conditions to equivalent transactions (similar to the "undue discrimination" principle in telecommunication licences) and tie-ins. These examples are of course not exhaustive.

Telespeed

Given the presence on the United Kingdom telecommunications market of an undertaking (*i.e.* BT) in a dominant position, it is possible that Article 86 may be of more than academic relevance in controlling and promoting competition within the United Kingdom. Indeed even whilst BT remained a state enterprise before 1984, there was ample demonstration of the applicability of Article 86 to PTTs and to BT in particular, in what is sometimes known as the Telespeed Case.[10] This case concerned the activities of a United Kingdom message forwarding agency called Telespeed. This company, in common with others, took advantage of the lower tariffs within the United Kingdom on traffic to North America in order to "hub" such traffic on the United Kingdom for reforwarding to and from the rest of Europe. The service involved reforwarding telex messages as well as reprocessing messages received in data form over data circuits and resending such messages as data to be received by terminals and read either by way of a visual display unit or printed out in hard copy.

Such activities appeared to be frowned upon by the ITU and more specifically the CCITT, whose recommendation F60 on telex operating methods exhorted PTTs to refuse to make telex service available to message forwarding agencies who were assisting in evasion of "the full charges due for the complete route." Presumably by "complete route" the CCITT were intending to refer to "direct" or "normal" route.

In implementation of this particular recommendation BT's predecessor, the Post Office, incorporated in its standard terms of service direct restrictions on reforwarding of messages.

In June 1979 Telespeed made a complaint to the Commission requesting the Commission to find that by introducing these restrictions BT had infringed Article 85(1) or Article 86. Significantly, Telespeed complained that the restriction also operated to prevent reforwarding even where there was no question of lower rates being charged or available than those applicable on direct routes from or to the originating or terminating countries.

Following the Commission's intervention, in October 1982 BT voluntarily agreed to amend its standard conditions to delete the offending restrictions (as to which see below) but the Commission nevertheless shortly afterwards issued its decision which found that Article 86 is both applicable to BT and had been infringed. The case was appealed by the Italian Government to the European Court which upheld the Commission decision. Much of the decision related to the status of both BT and the Post Office and applicability of the Treaty to a statutory monopoly. Whilst of interest in looking at the relevance of Article 86 to other European PTTs, the case's significance for telecommunications liberalisation is more in relation to its discussion of value added services and its conclusion that BT had gone further than was necessary to comply with CCITT Recommendations, which of themselves were not contrary to Article 86.

The Court highlighted the importance of the fact that the CCITT recommendation in question (F60) was directed only towards agencies which, "by means of abusive procedures" attempted to evade payment for the full charge due on the complete route. The Italian Government argued that the use of leased circuits for reforwarding messages was such an abuse, as was the use of special equipment to increase the volume and speed of messages. The Court found that there was no practical evidence of the use of leased circuits but commented no further; as regards the use of advanced equipment, the Court held this would not amount to an abuse as the Telespeed service simply took advantage of technical progress and benefited consumers generally.

Unfortunately the Court's judgment is somewhat rambling and unstructured in its style and it is not at all clear what its attitude would be in a case where it was clearly shown the agency was undercutting a PTT on price, through using leased circuits, and utilising multiplexers, concentrators and other technological enhancements. Normally up to the present time leased circuits have been charged for on the basis of a flat rate rental, but usage-sensitive tariffs for such circuits are beginning now to creep in. However, it is submitted that no matter how the tariffs are structured, and even where only a flat rate is payable, an attempt by a PTT to prevent or inhibit the use of leased circuits for value added services could, in many cases, be a contravention of Article 86. This might also involve, in effect, declaring the relevant CCITT recommendation itself to be against Article 86, or a concerted practice caught by Article 85.

It seems inevitable that the European Commission will some day have to return to a closer scrutiny of CCITT Recommendations, particularly in relation to competition by "resellers" and value added service providers.

Currently CCITT Recommendation F60 continues in the same form as was discussed in the Telespeed case. Meanwhile BT's current standard conditions of contract for telex service (Issue 1, August 1984) do indeed omit the particular restriction found by the Commission and the Court to offend Article 86. The same conditions, however, continue as previously certain restrictions on reforwarding of messages within the United Kingdom which originated outside the United Kingdom and reforwarding out of the United Kingdom messages originating in the United Kingdom. Both of these provisions would appear to curtail message reforwarding to and from originations and destinations within the United Kingdom and on the face therefore could yet affect inter-member state trade.

Customers of PTOs should be wary to ensure that PTO contract terms for use of telecommunication service are no more restrictive than the constraints imposed by United Kingdom licensing regulations, particularly in cases where these regulations are more liberal than, say, CCITT Recommendations. Hitherto, for example, BT's conditions of contract for private circuit service, in following the CCITT 'D' Recommendations, were more restrictive of the use of private circuits than the VADS licence. BT have now amended these conditions simply to require the customer to adhere to the terms of his applicable licence under the Act, the CBA or the WTA.

Telecommunications regulation from the point of view of the promotion and protection of competition can thus be seen to be an attempted amalgam of existing competition law enacted for trade and industry as a whole and new rules and regulations specific to telecommunications and mainly embodied in licences granted under the Act.

This is not a recipe for a wholly successful and consistent approach to the policing of the market-place and so it appears to be proving. In some areas real competition is helping to break down market barriers and market dominance and ensure better service, including contract terms, for customers; in others, such as the battle between BT and Mercury for the wallets and minds of customers for national switched service, the growth of effective competition has been helped by the Director General's determinations on interconnection,[11] but in certain matters he has no power to intervene either directly or with any immediacy. However, the Director General does have the ultimate power to take steps to seek modifications to PTO licences to try to plug any perceived gaps in the regime with which he has to work; it seems inevitable that one day soon a reference of some kind will be made to the MMC in order to achieve more effective competition regulation and to improve customer benefits.

Notes

[1] In Chaps. 4, at p. 59 and 6, at p. 105.
[2] S.18(7) of the Act.
[3] S.59(4) of the Act.
[4] See Chap. 3 at p. 38.
[5] S.50(2) of the 1973 Act.
[6] *Ibid.* s.64.
[7] Cmnd. 9715 (1986).
[8] To be given within six months, as specified by the Director General: Competition Act, s.7(6) and FTA, s.70.
[9] S.I. 1980, No. 979.
[10] Re British Telecommunications: *Italy* v. *European Commission* [1985] C.M.L.R. 368.
[11] See Chap. 4 above at p. 67.

CHAPTER 10

PROPERTY RIGHTS AND THE ENVIRONMENT*

A. INTRODUCTION

1. History

Before the passing of the Act the Post Office, and subsequently BT, had, like many other Government bodies, very little control imposed upon them, other than at common law, over their powers and duties so far as the acquisition of property and property rights were concerned. Furthermore, there was very little planning control imposed upon the Post Office in respect of telecommunications development. The privatisation of the telecommunications industry, and the change from a state monopoly to a commercial duopoly, has resulted in a significant body of statutory rights, duties and powers being enacted in order to regulate PTOs with regard to property rights and duties.

Since the Act came into force, there has been a general tightening up in planning control; nevertheless, the overriding principle still seems to be that development for telecommunications purposes should not be too strictly controlled, unless there are convincing reasons to the contrary. Conversely, as the powers exerciseable by a Government department have fallen away, they have had to be replaced by a statutory code of rights to permit telecommunications operators to acquire rights over, and interests in, land by compulsory means. Schedule 2 of the Act contains this code, which is referred to as "the Telecommunications Code."

2. Legislative Context

In considering the effect of the Act it is convenient to look at the subject in three main categories:

* This Chapter written by Stephen Kingsley.

(a) the rights and duties of PTOs concerning the public at large;
(b) the rights and duties of PTOs in respect of land in private ownership; and
(c) the rights of third parties either against PTOs or in respect of telecommunications facilities.

Whilst matters such as planning, rating, and street works are covered by separate legislation, rights against private landowners, and the rights of owners and occupiers of premises are contained in the Act itself.

It must be emphasised that the contents of this chapter deal with the additional rights granted to, and duties imposed upon, PTOs and that the normal rules of law relating to real property, and landlord and tenant matters, will still apply.

B. PUBLIC RIGHTS AND DUTIES

1. Planning and the Environment

Before the Act the Post Office, and subsequently BT, enjoyed considerable freedom from planning controls. Telecommunications installations erected pursuant to the Telegraph Acts 1863 to 1916 were not subject to planning control at all. Where the provisions of these Acts did not apply deemed planning consent was granted, as for other statutory undertakers, for certain developments, by Class XVIII of the Town and Country Planning General Development Order 1977 ("the General Development Order").

Since the Act came into force telecommunications development has become subject to planning control, although, as will be seen, there is still a considerable amount of development permitted under the General Development Order.

In addition to normal planning control, the terms of the relevant telecommunications Licences also impose obligations upon PTOs in respect of the environment. Schedule 4 to both the Licence issued to BT and the Licence issued to Mercury contain conditions to preserve the environment. For example, overhead telephone lines cannot be installed in Conservation Areas, except in specified circumstances; telecommunications apparatus cannot be installed above ground near listed buildings and ancient monuments without prior notice being given to the relevant planning authority and, if the planning authority object, then it can only be installed with the consent of the Secretary of State; there is a general Licence condition that overhead lines should be avoided where possible; and in National Parks, Areas of Outstanding Natural Beauty, and other similar locations, works cannot be carried out until written notice has been given to the relevant authority, except in the case of certain minor works.

2. Permitted Development

As mentioned above, a significant degree of planning consent is granted by the General Development Order, contained in Class XXIV and Class XXV. It should be noted that the previous rights granted to the Post Office and BT pursuant to Class XVIII have been withdrawn, and BT is no longer treated as being a statutory undertaker.

Class XXIV of the General Development Order permits the carrying out of development by a PTO on land occupied by the PTO and held either freehold or on a lease for a term of not less than ten years or in exercise of a right conferred on the PTO pursuant to the Telecommunications Code. The development permitted is in respect of the installation, alteration or replacement in, on, over or under any land (which is defined to include buildings) of any telecommunications apparatus where the telecommunications apparatus itself, unless installed upon a building or other structure, does not exceed a height of 15 metres above ground level. In the case of alteration or replacement, the new telecommunications apparatus must not exceed the height of the apparatus replaced, or the height of 15 metres, whichever is the greater.

In the case of the installation, alteration or replacement of telecommunications apparatus on a building or other structure:

(a) the height of the telecommunications apparatus taken by itself must not exceed ten metres, unless the building or other structure has a height of 30 metres or more, in which case the limit on the height of the telecommunications apparatus is 15 metres; and
(b) the highest part of the telecommunications apparatus when installed must not exceed the height of the highest part of the existing building or structure by more than ten metres in the case of a building which is 30 metres or more high; by more than eight metres in the case of a building which is more than 15 metres but less than 30 metres high; or by more than six metres in respect of a building which is of a height of less than 15 metres.

In the case of the installation, alteration or replacement of telecommunications apparatus other than a mast, antenna, public call box or any apparatus which does not project above the level of the surface of the ground, the ground or base area of the structure must not exceed 1.5 square metres.

However, it must be noted that in the case of the installation, alteration or replacement of a microwave antenna, or telecommunications apparatus which supports such antenna, on a building or other structure, the building or structure must be more than 15 metres in height, the size of the antenna must not exceed 1.3 metres in any dimension (but excluding any projecting feeder element) and the development must not result in the presence upon the building or structure of more than two microwave antennae. Also, the antenna, and any supporting structure, must, so far as practicable, be located so as to minimise its effect on the external appearance of the building.

In addition, Class XXIV permits the use of land in an

emergency for a period not exceeding six months for the stationing and operation of moveable telecommunications apparatus required for the replacement of unserviceable telecommunications apparatus, and the erection or placing of moveable structures on the land for the purposes of that use. Upon the expiry of the period of six months, or when the use is no longer required, the telecommunications apparatus must be removed and the land restored to its former condition.

Class XXV of the General Development Order grants planning consent for the installation, alteration or replacement on any building or other structure, other than a dwelling house, in circumstances other than those set out in Class XXIV (*i.e.* permanent development otherwise than by a PTO on its own land), of a microwave antenna and any structures intended for the support of such an antenna only. It is limited to cases where the building or other structure exceeds a height of 15 metres. In the case of a terrestial microwave antenna, under Class XXV the size of the antenna when measured in any dimension (but excluding any projecting feed element) must not exceed 1.3 metres and the highest part of the antenna or its supporting structure must not be more than three metres higher than the highest part of the existing building or structure upon which it is installed. In the case of a satellite antenna, under Class XXV the size of the antenna taken together with its supporting structure (but excluding any projecting feed element) must not exceed 90 centimetres. No more than two microwave antennae may be erected on a building or structure at any time, without the grant of specific planning consent. Again, Class XXV contains a condition that the antenna should be sited so as to minimise its effect on the external appearance of the building or structure upon which it is erected, and a further condition that when an antenna is no longer required it should be removed.

It will be noted that Class XXIV is exclusively for the benefit of PTOs and Class XXV is for the benefit of a PTO carrying out development otherwise than on its own land

and for the benefit of a private individual or body, who is either connecting into a telecommunications system by means of microwave, or else requires a microwave antenna for some purpose unconnected with a public telecommunications system.

In respect of dwelling houses, Class I of the General Development Order permits the installation, alteration or replacement of a single satellite antenna upon the house itself or within the curtilage of the house. This is subject to the proviso that the size of the antenna should not exceed 90 centimetres when measured in any direction (but excluding any projecting feed element) and that the highest part of the antenna should not be higher than the highest part of the roof of any building on which it is erected. A satellite antenna cannot be installed in a position beyond the forwardmost part of any wall of the original dwelling house which fronts onto a highway.

3. Exceptions and Limitations to Permitted Development

It is always open to a local planning authority to restrict or entirely remove the deemed planning consent by virtue of its rights contained in Article 4 of the General Development Order. In addition, the Town and County Planning (National Parks, Areas of Outstanding Natural Beauty, and Conservation Areas, etc.) Special Development Order 1985 provides that Class XXV of the General Development Order shall not apply to land within National Parks, Areas of Outstanding Natural Beauty, Conservation Areas and other specified areas of scenic or natural interest.

The Special Development Order also makes Class XXIV of the General Development Order subject, in such areas, to an additional condition and an additional limitation in that not less that eight weeks' notice (except in case of emergency) must be given to the local planning authority of an intention to carry out permitted development and that the provisions of Class XXIV shall not apply in respect of the installation or alteration of a microwave antenna, or any apparatus which includes or is intended for the support of such antenna unless this is merely a replacement of such antenna with one of the same size, design, appearance and location.

4. Government Advice on Planning Policy

Having provided that, other than development permitted by the General Development Order, telecommunications development is to be subject to planning control, the Secretary of State for the Environment states (in Circular No. 16/85) that it is Government policy "to facilitate and encourage the growth of telecommunications" and that "the planning system should encourage, and not place any unnecessary obstacles in the way of, development in this field".

The Secretary of State then goes on to say that owing to "the special needs and technical problems of telecommunications development . . . these will necessarily have to prevail over normal planning policies". Furthermore "authorities should not question the need for the service which a proposed development is to provide, or the system by which the service is to be provided, or seek to prevent competition between different operators". Planning authorities cannot question development on grounds of radio interference unless there is "firm evidence that significant and irremedial radio interference with other electrical equipment of any kind is a probability" and only if there is "already clear evidence that significant radio interference would arise, or will probably arise, and that no practical remedy is available, will there be any justification for taking it into account in reaching a decision". In respect of radiation hazards, the Secretary of State has stated that the power output of radio installations, including those using microwave frequencies, are well within internationally accepted standards and that "other than in the most exceptional circumstances there is no reason for planning authorities to take such issues into account".

In January 1988 the Department of the Environment issued a Planning Policy Guidance note (PPG 8) upon the subject of telecommunications. It repeats the advice contained in Circular 16/85 and also contains some useful appendices on telecommunications systems, permitted development and regulatory controls on PTOs.

5. Rating

In addition to normal property and buildings occupied by telecommunications operators, the network of cables, wires, ducts and radio antennae making up the telecommunications systems of PTOs are subject to rating. Section 31 of the Act amends the Local Government Act 1974 to apply to the rating of the systems of all PTOs, in a similar way to other public utilities. For public utilities in general, such networks of pipes and ducts, etc. have been rated nationally, and the resulting figure divided between local rating authorities, based upon the profits of the public utility concerned. It should be noted that cables themselves are not rateable, pursuant to section 21 of the General Rate Act 1967 and the Plant and Machinery (Rating) Order 1960, although the ducts in which they are laid, and the space the cables occupy, are subject to rating.

However, the profits basis of assessment does not apply so far as BT is concerned, as an Order has been made under section 19 of the Local Government Act 1974 which has the effect of deeming that BT shall not be treated as a public utility for such purpose and that its telecommunications system is to be rated by means of a special formula (by reference to numbers of exchange lines) for the entire network, again apportioned between the different rating districts.

So far as Mercury is concerned, no formula has yet been promulgated, but it is believed that such a formula may well apply in the future. Until such time as an Order is made under section 19 of the Local Government Act 1974, the telecommunications system of Mercury will be rated as for any other public utility undertaking.

6. Street Works

By virtue of paragraph 9 of the Telecommunications Code PTOs are subject to the provisions of the Public Utilities Street Works Act 1950. There are thus obligations to consult with other public utilities, as well as with the highway authority, prior to carrying out works in the public highway. In addition, there are further obligations contained in the relevant Licences so that BT and Mercury have to give notice to the highway authority before carrying out any works and must consider any written representations made by the highway authority within eight days of the giving of such notice, in the case of underground service lines or overhead lines, and within 29 days in any other case. Works cannot commence until the relevant period has expired, unless the highway authority otherwise agree.

The Licence conditions also say that new cables laid underground must be laid in ducts and that lines installed over a carriageway should be at height of not less than 5.5 metres above the carriageway, or 6.5 metres in the case of a designated high load route, unless the highway authority otherwise agree.

7. Works by other Public Utilities

If a local authority, railway or transport undertaking or another public utility proposes to carry out works to land which may involve the permanent or temporary alteration of telecommunication apparatus, paragraph 23 of the Telecommunications Code lays down the procedure to be followed. Not less than ten days' notice must be given before commencing works, except in the case of emergency. The PTO then has the choice as to whether to carry out the works itself or whether merely to supervise the carrying out of the works by the third party. Any expenses incurred by the PTO are to be borne by the third party. Any works carried out by the third party, whether or not actually supervised by the PTO, must be carried out to the satisfaction of the PTO.

8. Land Registration

Paragraph 2(7) of the Telecommunications Code provides that where rights are granted to a PTO pursuant to paragraph 2 or paragraph 3 of the Telecommunications Code these rights are specifically excluded from any requirement for registration of interests in, charges on, or other obligations affecting land.

C. PRIVATE RIGHTS AND DUTIES

1. Compulsory Purchase

Section 34 of the Act provides that a PTO may be authorised by the Secretary of State to acquire land compulsorily if it is required for, or in connection with, the establishment or running of the PTO's system. The Acquisition of Land Act 1981 will apply as if the PTO were a local authority within the meaning of that Act. Before the making of an Order the consent of the Director General is required. In addition to acquiring land itself, there is included the right to compulsorily acquire an easement or other right over land by the creation of a new right.

2. Compulsory Grant of Rights

Under the Telecommunications Code the consent of a landowner or occupier is required before the erection of telecommunications apparatus (paragraph 2 of the Telecommunications Code) or the obstruction of access (paragraph 3 of the Telecommunications Code). If a landowner refuses to co-operate, a PTO may apply to the Court under Paragraph 5 of the Telecommunications Code for an Order dispensing with the need for the agreement of the landowner in question. The PTO must give written notice to the person in question and if, within a period of 28 days, no consent is forthcoming, the application can then proceed to the Court.

The Court is to make an Order if it is satisfied that any prejudice caused by the Order is capable of being adequately compensated for by money or is out-weighed by the benefit accruing from the Order to the person whose access to a telecommunications system will be secured by the making of the Order. In determining the extent of the prejudice, and the weight of the benefit, the Court is to have regard to all the circumstances and to the principle that no person should unreasonably be denied access to a telecommunications system. The Court has, pursuant to paragraph 7 of the Telecommunications Code, the discretion to fix financial terms in such circumstances.

It should be noted that the Court of appropriate jurisdiction, in respect of all matters referred to in the Telecommunications Code, is the County Court.

Paragraphs 10 and 12 of the Telecommunications Code also grant rights to a PTO to lay lines across any land, subject to giving 28 days' notice, except in emergency. In the event that objection is made by a landowner following the service of notice compensation is payable. Overhead lines must not interfere with any business carried on upon the land affected.

If a tree overhangs any street and in doing so interferes or obstructs the working of any telecommunications apparatus, or will interfere or obstruct such apparatus which is about to be installed, the PTO may, pursuant to paragraph 19 of the Telecommunications Code, require such trees to be lopped so as to prevent an obstruction or interference. The landowner has 28 days to serve counter-notice objecting to the lopping of the tree and, in which event, the notice shall only have effect if confirmed by an Order of the Court. If the landowner does not comply with a notice served by the PTO then the PTO may lop the tree in question.

It should be noted that paragraph 21 of the Telecommunications Code provides that, once telecommunications apparatus has been installed upon any property, the PTO cannot be required to remove the apparatus (notwithstanding any agreement to the contrary) unless the person requiring the removal of the apparatus obtains an Order of the Court to enforce such removal.

Paragraph 26 of the Telecommunications Code specifically states that the provisions of the Telecommunications Code shall apply to Crown Land, in addition to land in private ownership.

3. Restrictive Covenants

One important provision of the Telecommunications Code is in respect of restrictive covenants. Where a PTO exercises a right conferred pursuant to paragraph 2 or paragraph 3 of the Telecommunications Code it is deemed to be doing so in exercise of a statutory power and this will thus over-reach a restrictive covenant affecting the land unless that covenant was itself entered into pursuant to any enactment (paragraph 4 of the Telecommunications Code). A covenant entered into with a local authority pursuant to section 52 of the Town and Country Planning Act 1971 will continue to bind a PTO, but a covenant entered into by an individual on the purchase of land will not bind the PTO.

4. Duties and Obligations

Where rights are acquired by a PTO pursuant to paragraph 2 or paragraph 3 of the Telecommunications Code compensation may be payable to a landowner under paragraph 4 of the Telecommunications Code. In addition, paragraph 16 of the Telecommunications Code gives rights of compensation for injurious affection to the owners of neighbouring land to that upon which the PTO has exercised rights.

If a PTO installs overhead apparatus then, under paragraph 18 of the Telecommunications Code, there is an obligation upon the PTO to affix a notice to the apparatus giving an address for the making of objections in respect of such installation.

D. RIGHTS OF THIRD PARTIES

1. Rights of potential subscribers against Lessors

Section 96 of the Act (which has yet to be brought into force) grants rights to an individual, who wishes to install telecommunications equipment inside a building for purposes connected with the provision to that person by a PTO of a telecommunications service, to have any prohibition or restriction contained in the lease under which he holds his interest in the property treated as being subject to a provision that his lessor is not to withhold consent unreasonably. Furthermore, where a provision of a lease imposes (whether by virtue of section 96 or otherwise) a requirement on the lessor under a lease not to withhold his consent unreasonably, the question of whether consent is unreasonably withheld is to be determined with regard to all the circumstances and to the principle that no person should unreasonably be denied access to a telecommunications system. This section only applies to a lease for a term of a year or more granted on or after the day on which the section comes into force.

2. Rights of potential subscribers against third parties

If a potential subscriber to a public telecommunications system is unable to obtain access to such system because the agreement of a third party is required to confer any right over the land belonging to such a third party and such agreement is not forthcoming, the potential subscriber is granted certain rights to apply to the Court by paragraph 8 of the Telecommunications Code. If no notice has been served by the PTO pursuant to paragraph 5 of the Telecommunications Code, the potential subscriber may serve notice upon the PTO requiring such notice to be served and, if the PTO does not serve notice within 28 days, then the potential subscriber may, in turn, serve notice on behalf of the PTO on the third party. If the PTO has given notice pursuant to paragraph 5 of the Telecommunications Code, or if the potential subscriber has served notice on the PTO's behalf, and the PTO has not yet commenced proceedings in the Court, then the potential subscriber may himself commence proceedings on behalf of the PTO.

3. Rights against PTOs

Paragraph 17 of the Telecommunications Code gives the right to object to the installation of telecommunications apparatus which is at a height of three metres or more above ground at any time within the period of three months, beginning with the completion of the installation, unless this is merely the replacement of existing apparatus. The person serving the notice of objection may apply, not earlier than two months nor later than four months from the giving of the notice, to the Court to have the objection upheld.

If the apparatus appears to the Court materially to prejudice the objector's enjoyment of an interest in the land in respect of which the objection is made then, unless the cost of providing the service will be substantially increased, or the quality of the service provided substantially diminished, or unless it involves the PTO in substantial additional expenditure, the Court will uphold the objection and the Court may either direct the alteration of the apparatus, or authorise the installation of substitute apparatus.

Paragraph 20 the Telecommunications Code gives the right to any person with an interest in land to require a PTO to alter apparatus in order to carry out a proposed improvement of the land in which he has an interest. If, within 28 days of giving notice to such effect the PTO does not serve counter-notice, stating that it does not intend to make such alteration, it must comply with the request. If counter-notice is served, the PTO is only required to make the alteration pursuant to an Order of the Court. The cost of making any alteration is to be borne by the person making the request.

It should be noted that whilst the majority of the provisions of the Telecommunications Code can be excluded by specific agreement to the contrary, this does not apply to the provisions of paragraph 8(5) of the Telecommunications Code, which makes void a covenant, condition or agreement which would have the effect of preventing or restricting the taking of any person as a potential subscriber to a telecommunications system, nor to the provisions of paragraph 21 of the Telecommunications Code, which deals with the right of a PTO to refuse to remove or alter apparatus save pursuant to an Order of the Court.

E. CONCLUDING NOTE: LEGAL AND JUDICIAL CONTEXT

Telecommunications development is affected by a multiplicity of existing laws and regulations, in addition to the legislation that has come into effect since 1984, and this chapter does not attempt to deal exhaustively with those aspects of telecommunications law relating to property. Eventually, as the law develops and judicial decisions have been made as to the interpretation of the various statutory provisions which relate to this very new area of law, the subject may well merit an entire book of its own. Those aspects of existing law, such as planning and rating, are already dealt with in some depth in other works, although the particular impact of the recent legislation may perhaps not yet be fully absorbed. In the short period since the Act came into force there has been very little opportunity for the Court to consider the effect of the Telecommunications Code and until such time as this occurs the only valid authority is the Telecommunications Code itself.

CHAPTER 11

LEGAL ISSUES

A. COPYRIGHT

1. General

The provision of conventional telecommunication services in itself has few copyright ramifications. However, providers of basic telecommunication services have increasingly moved from their origins in the simple transmission of private messages to permitting their facilities to be used for public communications and so have become carriers of services with copyright ramifications, such as cable and satellite broadcasts. This is however not the place for a detailed analysis of the copyright aspects of being a service provider, and the discussion in this section is primarily directed to the position of PTOs and other telecommunication service providers.

Copyright in the United Kingdom has for the last 30 years been governed by the Copyright Act 1956, as successively amended. Significant amendments from the point of view of cable and satellite broadcasting were introduced by the CBA. Copyright law is now being totally restated in the Copyright Designs and Patents Bill ("the Bill") currently before Parliament, and which is used as the basis for this discussion.

As in the past, the Bill provides that copyright subsists in various types of works, including:

(a) original literary, dramatic, musical or artistic works; and
(b) sound recordings, films and broadcasts or cable programmes.

Copyright in such works can be infringed in a number of ways. For the present discussion, the relevant infringing acts are and will under the new law be the following ones, if unauthorised by the copyright owner:

(a) copying (which includes storing the work in a computer)
(b) performing, showing or playing the work in public
(c) broadcasting the work or including it in a cable programme service.

Thus the mere transmission or (to use the regulatory term "conveyance") of a message from one point to another would not normally enter the realms of copyright unless this act constituted the broadcasting of a copyright work its public performance or its inclusion in a cable programme service, or the message was, at some stage, stored in a computer, even if only momentarily.

As with all discussions of copyright, it should be remembered that copyright in a broadcast or a cable programme is quite distinct and separate from the various copyrights, for example musical and sound recording copyrights, in the actual material broadcast or included in a cable programme service.

2. Position of telecommunication operators

The position of PTOs (for example long-line PTOs and cable television operators), as carriers conveying signals for ultimate presentation to the public, is not spelled out in the Bill since generally speaking their activities are not such as to confer any copyrights on them or to expose them to liability for copyright infringement. Rather, their core activities constitute exclusions to the definitions of "broadcasting" and "cable programme service" in the Bill and in so far as they are involved in providing telecommunication services to broadcasters and cable programme service providers they are absolved from liability. The terms "broadcast" and "cable programme services" are discussed in more detail in section 3 below.

Telecommunications operators avoid liability as mere carriers of broadcasts or cable programme services by virtue of the provisions in clauses 6(3) and 7(5) respectively of the Bill. The first provides that the person "making the broadcast, broadcasting a work, or including a work in a broadcast" (and so attracting liability for infringement, but also receiving the benefit of owning copyright in its transmission) is the person who has responsibility for the contents of the broadcast; this excludes the public telecommunications operator in its capacity as a mere carrier. The second clause provides that the person "including a cable programme or a work in a cable programme service" (and so attracting liability for infringement and the benefit of copyright protection) is "the person providing the cable programme service in question," which also serves (although somewhat less clearly) to exclude the mere operators of the cable system from liability as carriers.

So far this section has considered liability as a PTO/carrier for infringement by unauthorised broadcasting or inclusion in a cable programme service. Two other potential infringement liabilities arise—one in connection with "performing, showing or playing the work in public" and one with "copying the work." As to the first of these, that of performing, showing or playing the work in public, clause 19(4) of the Bill provides that where such infringement takes place "by means of apparatus for receiving visual images or sounds conveyed by electronic means, the person by whom such visual images or sounds are sent" is not regarded as responsible for the infringement. Whilst this primarily serves to exclude from liability actual broadcasters and providers of cable programme services it inevitably also excludes the mere carriers of such services.

Outside of broadcasting, the increasingly computer-controlled operation of telecommunication networks should mean, in theory at least, that the same copyright issues which have become apparent in relation to the law on data processing activities will be equally relevant to telecommunication. This will be most evident in the case of electronic mail, packet switching and other value added services which utilise computers to store, reproduce, bundle, segregate and redirect messages. As mentioned above in the provision of enhanced telecommunication services the storage of a "work" in a computer could be involved: this would be copying of the work comprised in the message.[1] In such circumstances the customer receiving a value added service which involves any such deemed copying of his work can probably be said to have impliedly granted the system operator/service provider a licence to perform any necessary copying of the message as part of the contracted service. It may even come about that such operators and service providers will include appropriate provisions in their conditions of service in order to put the matter beyond doubt.

3. Broadcasts and Cable Programmes

At the time of writing the law of copyright as to broadcasts and cable programmes is governed by sections 14 and 14A of the Copyright Act. However, these will shortly be superseded by clause 20 of the Bill which gives the owners of copyright works the right to authorise or prohibit broadcasting of the work or its inclusion in a cable programme service.

Clause 6 of the Bill defines a "broadcast" as meaning a transmission by wireless telegraphy[2] of visual images, sounds or other information which:

(a) is broadcast for general reception; or
(b) is transmitted for presentation to members of the public; or
(c) is a satellite transmission which is capable of being lawfully received by members of the public and either is *not* encrypted or is encrypted but appropriate decoding equipment is generally available to the public from the broadcaster or with his permission.

A "cable programme" is defined in clause 7 of the Bill as being any item included in a cable programme service, whilst a "cable programme service" is defined in the same clause as being a service involving the sending of visual images, sounds or other information by means of a telecommunications system (excluding wireless telegraphy) for reception either at two or more places or for presentation to members of the public. Such a telecommunication system would of course have to include a cable of some kind. Cable programme services exclude the following:

(i) what are essentially "interactive" services;
(ii) services run for the purposes of a business where the telecommunication system over which the service is provided is not connected to any other telecommunication system;
(iii) services run by a single individual for his own domestic purposes and again where the telecommunication system involved is not connected with any other system;
(iv) services where all the apparatus comprised in the relevant telecommunication system is situated in or connects premises in single occupation and again where that telecommunication system is not connected to any other system; and
(v) services run for broadcasters or persons providing cable programme services or programmes for such services.

As was confirmed by the Government in the course of the Committee Stage of the Bill in the House of Lords[3] the intent of this rather complex collection of provisions was to exclude from the definition of cable programme services activities such as "ordinary telephone conversations, teleshopping, telebanking or video conferencing services; all of which are communications of an essentially private kind."

Under clause 9 of the Bill, the author and first owner of the copyright in a broadcast is the person making the broadcast and, in the case of a cable programme, is the person providing the cable programme service in which the programme is included. However, copyright will not arise in a cable programme which is included in a cable programme service by reception and immediate retransmission of a broadcast.

Cable Programme Services: Reception and Retransmission

Under clause 65 of the Bill, copyright in a television broadcast or sound broadcast (as well as in the work included in the broadcast) made from the United Kingdom will not be infringed where the broadcast is received and immediately retransmitted for inclusion in a cable programme service either:

(a) because of the Cable Authority's duty to secure the inclusion of certain programmes in cable services (under the CBA a franchise is not granted to a Cable Television operator unless the operator carries the four channels of the BBC and IBA); or
(b) where and to the extent that the broadcast is made for reception in the area in which the cable programme service is provided and it is not a satellite transmission nor an encrypted teletext transmission. In the latter case such "in-area" cable retransmission is duplicating the broadcaster's direct reception audience and the copyright owner does not need a second bite of the cherry, having already settled with the broadcaster.

4. Satellite Broadcasting

As has been seen in 1.3 above "broadcasting" includes certain satellite transmissions. Satellite transmission of copyright material has been complicated by the different types of communication satellite, which essentially are as follows:

(a) Direct Broadcasting Satellite ("DBS"): a high powered satellite transmitting over a wide area (known as the "footprint") for direct reception by the public without the interposition of a receiving station or of cable distribution. With pay television the broadcast messages would have to be encrypted and subscribers to the programme services would have a suitable decoder installed in or plugged into their television sets. DBS transmissions will, initially at least, only be receivable by use of SMATV dish aerials.

(b) Fixed Service Satellite ("FSS"): a low powered point-to-point satellite whereby television programmes are transmitted to a ground (receiving) station for rebroadcasting or distribution by cable.

Direct Broadcasting Satellite ("DBS")

DBS involves an "uplink" to the satellite and a "downlink" from that satellite to the ground. The uplink involves the transmission of programmes direct to the satellite; at first sight such transmission would appear not to involve broadcasting for direct reception by the general public. The broadcasting would be effected by the downlink transmission only, which in any event might be said to originate outside the United Kingdom jurisdiction. The CBA therefore amended the Copyright Act to ensure that DBS broadcasts did not escape copyright liability on these grounds. These provisions are now reproduced in clause 6(4) of the Bill.

As with terrestrial telecommunication sections 14 and 14A of the Copyright Act, gave copyright owners the right to authorise or prohibit public broadcasting of their works by DBS. This is reaffirmed by clause 20 of the Bill.

Fixed Service Satellite

FSS transmissions have not until recently been regarded as a communication to the public and therefore as broadcasting, and providers of FSS programmes have not been under any strict obligation to obtain authorisation or pay royalties to copyright owners. This position flowed logically from the fact that generally speaking FSS were not intended for reception directly by the general public but only indirectly by the intermediary of cable operators. However, in practice FSS transmissions can be picked up by SMATV and therefore constitute communication of copyright works to the public.

As mentioned above (1.3) this anomaly is being corrected in the Bill, by its defining a broadcast as not only one which is intended for general reception but also one "accessible" (*i.e.* capable of being received) by any member of the public with the appropriate (for example decoding) equipment and any necessary licence for that purpose.

5. International Broadcasting

Subject to the qualifications and proposed legislation mentioned above, in the United Kingdom satellite transmission, both by DBS and FSS, constitutes or will constitute broadcasting activity. However, since the satellite is merely a receiver/transmitter in space it follows logically that it is the broadcasting organisation which first beamed the programme up towards the satellite which is to be treated as the broadcaster in the country of reception. This is confirmed by clause 6(4) of the Bill which states that "the place from which a broadcast is made is, in the case of a satellite transmission, the place from which the signals carrying the broadcast are transmitted to the satellite." It is therefore the broadcaster's national copyright legislation and any applicable international convention law which will be relevant to the position.

6. EEC: Cross-Frontier Broadcasting

In the first Coditel case[4] the European Court held, in relation to a reference on Articles 59 and 60 of the Rome Treaty, that an exclusive licensee in Belgium in respect of rights in a film was able to rely on his right to prohibit cable distribution in that country of a German broadcast incorporating the film in question. As a result of the Coditel case, the European Commission became concerned about the restrictive effects of copyright on cross-frontier broadcasting, particularly re-transmission of foreign programmes by cable. In its examination of the problem in its EEC Green Paper of June 1984 "Television without Frontiers"[5] the Commission therefore proposed that the right to authorise or prohibit the cable retransmission of broadcasts contained in legislation of various Member States should be repealed and replaced by a form of compulsory licensing procedure with royalties collectable through collection societies. The Green Paper came under tremendous criticism in this respect, particularly in the United Kingdom, and the Council has now issued a draft directive adopting a slightly more restrained approach.

So far as this draft EEC directive concerns copyright, this book is not the place for detailed analysis of all its provisions, which extend not just to cross-frontier broadcasting but also to programme production, advertising and the protection of children. It suffices to state that the directive proposes that the Member States should require cable operators, who retransmit broadcasts, to have suitable prior agreements with the copyright owners for this purpose. More controversially, however, it also proposes a statutory licence and "equitable remuneration" procedure for legitimising such retransmissions where such agreements cannot be negotiated.

B. DEFAMATION

Under the Defamation Act[6] the broadcasting of words by means of wireless telegraphy must be treated as publication in a permanent form and therefore capable of being libellous.

Under section 28 CBA, the publication of words in the course of a programme included in a cable programme service is also to be treated as publication in a permanent form.

C. CRIMINAL LIABILITY

1. Interception of Communications

The Interception of Communications Act 1985 is the all-embracing legislative context. This Act was designed essentially to do two things. First to create a statutory framework for the official authorisation of interception of communications sent not only through a public telecommunication system ("telephone tapping") but also through the post; secondly, to create an offence of intercepting such communications without authorisation and to provide remedies against improperly authorised interception.

Section 1 creates the offence of intentionally intercepting communications in the course of their transmission by means of a public telecommunication system, subject to four exceptions, these being:

(i) where the communication is intercepted "in obedience to a warrant" issued under section 2;
(ii) where the person intercepting has reasonable grounds for believing that the person to whom, or the person by whom, the communication is sent has consented to the interception;
(iii) where the communication is intercepted "for purposes connected with the provision of . . . public telecommunication services";
(iv) where the communication is by wireless telegraphy and is intercepted, with the authority of the Secretary of State, "for purposes in connection with the issue of licences" under the Wireless Telegraphy Act 1949 or the prevention or detection of interference with wireless telegraphy.

It should not be thought that this Act is about "bugging." The word "intercept" is not defined in the Act but its dictionary definition (for example O.E.D.—to prevent, check, stop or hinder) would normally apply; bugging and other forms of surveillance or eavesdropping by devices physically located at either "end" of a communication would not necessarily involve interception as such. There is clearly an argument that where the eavesdropping does not involve physical interference but simply monitoring or "listening in", this does not constitute interception. The point would arise particularly with respect to wireless telegraphy transmissions, such as by a cellular radio telephone system. Yet where the eavesdropper has thus successfully "picked up" a message in the course of its communication, he has to all intents and purposes intercepted it.

In passing it can be noted that in December 1984 the Secretary of State for the Home Department published strengthened guidelines to the police on the use of surveillance devices.[7]

Section 2 enables the Secretary of State to issue warrants for the interception of communications and for the disclosure of intercepted material. No warrant may be issued, however, unless the Secretary of State considers that the warrant is necessary:

(a) in the interest of national security;
(b) for the purpose of preventing or detecting serious crime; or
(c) for the purpose of safeguarding the economic wellbeing of the United Kingdom.

2. Offences by PTO Employees

Under the Act (as modified by the Interception of Communications Act) there are separate offences concerning modification, interception and disclosure of messages perpetrated by employees of public telecommunication operators. A person engaged in the running of a public telecommunication system who "otherwise than in the course of his duty intentionally modifies or interferes with the contents of a message sent by means of that system" is guilty of an offence (section 44). Likewise a person who, again otherwise than in the course of his duty intentionally discloses to any person the contents of any message which has been intercepted in the course of its transmission by means of that system or discloses any information concerning the use made of telecommunication services provided for any other person by means of that system is also guilty of an offence (section 45). In relation to the latter section there are exceptions provided for disclosures made in connection with the prevention or detection of crime or criminal proceedings, for disclosures under a warrant issued by the Secretary of State under the Interception of the Communications Act and for any disclosure as to the use made of telecommunication service where this is in the interest of national security or in pursuance of an order of the Court.

Unfortunately, section 45 will not apply to catch a PTO employee who wrongfully *uses* information comprised in a message which has been intercepted in the course of its transmission.

3. Fraudulent and Improper Use of Telecommunication Systems

One of the more notorious features of computer law has been its preoccupation with "hackers" and "hacking" and how this is to be approached under criminal law.

"Hacking," if viewed simply as unauthorised entry, a telecommunication system is itself possible, for example by the misuse of an authorisation code for access to an electronic mail system. A person engaging in such activity would be likely to commit an offence under section 42 of the Telecommunications Act. That section applies to a person who "dishonestly obtains a service provided by means of a licensed telecommunication system with intent to avoid payment of any charge applicable to the provision of the service."

In the Case of *R.* v. *Gold & Schifreen*, computer hackers were prosecuted for unauthorised access to the BT Prestel Computer Network. The prosecution was brought under the Forgery and Counterfeiting Act 1981 on the grounds that the accused had created a "false instrument" namely "a device on or in which information is recorded or stored by electronic means with the intention of using it to induce the Prestel Computer to accept it as genuine . . ."

At first instance the accused were convicted but these convictions were overturned on appeal. Lord Chief Justice, Lord Lane, considered that the "false instrument" would either have to be electronic impulses representing the Customer Identification Number and password falsely provided or the "user segment" being that part of the Prestel Computer receiving information keyed into it and storing that information simply for the purpose of its verification. He then held that neither of these things constituted a disc, tape or sound track or any other device on which information was recorded or stored and accordingly there had been no false instrument.

It seems surprising that the prosecution was not brought under section 42 of the Act. Undoubtedly a service (which for these purposes can be any kind of service) had been obtained dishonestly. The Prestel Computer Network must have been embodied within or connected to a telecommunication system in order that its messages could be conveyed from it into the BT system and thence to the hacker.

A more straightforward application of a criminal act caught by section 42 would clearly be the misuse of a public phonebox in order to obtain service free of charge.

Control over pornographic and other offensive messages sent via telecommunication systems is provided by section 43 of the same Act whereby any person who sends a message that is grossly offensive or of an indecent, obscene or menacing character, or sends a false message for the purpose of causing annoyance, inconvenience or needless anxiety to someone is also guilty of an offence.

Where obscene or offensive messages are sent as part of an "information" or "entertainment" service provided at premium rates, the matter could come before the Committee set up to interpret the ATIEP Code of Practice.[8]

4. Fraudulent Reception of Broadcasts and Cable Programmes

"Signal piracy" as it is called, is to be dealt with in clauses 269 to 271 of the Bill. Essentially, anyone who dishonestly receives a programme included in a broadcast or cable programme service, with intent to avoid payment of any charge for its reception, is guilty of an offence (clause 269). In addition the broadcaster or cable operator or other person charging for the service has the same rights and remedies against sellers and importers, and the same rights of seisure or delivery up of apparatus designed to assist such dishonest receipt, as a copyright owner has in respect of infringers and infringing articles (clause 270).

5. Wireless Telegraphy: Misleading Messages and Interception and Disclosure of Messages

In relation to the misuse of wireless telegraphy there are offences very similar to those mentioned in the earlier parts of this Chapter.

Under the WTA (section 5) any person who by means of wireless telegraphy sends any message which is to his knowledge false and misleading and is to his knowledge likely to "prejudice the efficiency of any safety of life service or endanger the safety of any person or of any vessel aircraft or vehicle" is guilty of an offence. Similarly, any person who otherwise than under the authority of the [Secretary of State] or in the course of his duty as a Crown Servant either uses any wireless telegraphy apparatus with intent to obtain information as to the contents, sender or addressee of any message (whether sent by means of wireless telegraphy or not) which he is not authorised by the [Secretary of State] to receive or who, except in the course of legal proceedings discloses any information as to the contents, sender or addressee of any such message [being information which would not otherwise have come to his notice] is also guilty of an offence.

D. CONTRACTUAL LIABILITY

1. Background

The whole area of legal liability in relation to the provision of telecommunication services is very new, essentially because prior to 1984 BT's liability in tort for failures of performance was limited by statute and its terms of service were embodied in "schemes" rather than in contracts. Thus BT could not be held liable for loss or damage with respect to any default in providing service.

This privilege was removed on August 5, 1984 when the relevant provisions of the Act came into force.

All PTOs now provide their services under what should be published terms and conditions filed with OFTEL pursuant to their licences. In BT's case only there is one qualification to be made to this: its terms of service for persons who became its customers before August 5, 1984 are (under Schedule 5 paragraph 12 of the Act) "deemed" contracts, these terms being embodied in the Telecommunications Scheme 1984 but otherwise being essentially the same as BT's standard conditions of contract.

2. Service Standards: Liability and Limitations

In the absence of anything contained in a contract to the contrary, the supply of telecommunication services falls squarely within the Supply of Goods and Services Act 1982. The effect of this is that under section 13 of that Act there is an implied term that the supplier shall exercise reasonable care and skill in the supply of the services. This is not a service warranty which is particularly apt for or susceptible to the vagaries of telecommunication, particularly a modern digital network where machines are more likely to be involved and occasionally fail to function than are people. However, since BT's network is still predominantly analogue in its technology and is only gradually converting to digital, the scope for faults in the service is still considerable and, certainly so far as switched service is concerned, it can be expected that not only BT but other PTOs reliant upon BT for interconnection will continue to shirk specifying any particular standard of service. In the absence of any exclusion and limitation on clauses, the liability of the PTOs with respect to any failure in performance of their services could be quite open-ended. The principles in *Hadley* v. *Baxendale*[10] apply to a telecommunication service contract as to any other contract. Under the "first rule" in this case all such damages as are the natural consequence of the breach are recoverable; under the "second rule" extraordinary losses are recoverable where these were in the contemplation of the parties as a probable result of the breach: *Victoria Laundry (Windsor) Ltd.* v. *Newman Industries Ltd.*[11]

Applying these considerations to telecommunication services, it must frequently be known to a PTO, particularly in relation to large orders from equally large customers, such as banks, how the customer intends to put the telecommunication service to use and the types of message that will be carried, for example whether this includes batches of data and whether that data is sensitive in any way, perhaps in comprising confidential or financial information. If such a customer suffers any loss as a result of a breakdown in the telecommunication service the effects could be immediate and irreparable and yet be purely financial in their impact. Accordingly, a PTO could incur such liability, whether consequential or direct. It may be that for these reasons, as well as general commercial practice, that PTOs have adopted the practice in their standard conditions of service of excluding and limiting their liability quite substantially. BT's standard conditions state:

> "Service is not fault-free and the customer shall be entitled only to the quality of service provided by BT from time to time for its telephone service to customers generally."

BT's conditions go on to exclude liability for loss of profits, business and all kinds of consequential loss and limits its liability for one incident to £1 million and, for any series of incidents arising in a 12 month period, to £2 million. Mercury has similar limitations of liability in its standard switched services contracts.

It is of course not to be forgotten that these exclusions and limitations are subject to the "reasonableness test" in the Unfair Contract Terms Act 1977 on two counts. This test applies, in the case of breach of contract, where the PTO is dealing with a "consumer"[12] or where the PTO is operating on the basis of standard terms and conditions (which it would normally do with all its customers, whether as consumers or in business).

The reasonableness test also applies where the PTO is guilty of negligence. Loss or damage arising from negligence, other than personal injury or death, is subject to this test. However, as is well known, under section 2(1) of the UCTA that it is not possible at all to exclude or restrict liability in negligence for personal injury or death by reference to any contract term.

3. Regulatory Modifications to Contractual Liability

It should not be forgotten that there are occasionally examples of regulatory underpinning to contractual obligations in relation to telecommunication services.

Under Condition 1 of BT's Licence which contains its "universal service obligation" discussed above,[13] BT is required to provide voice telephony and certain other telecommunications services to every person who requests them in any place in the United Kingdom (other than Hull), unless the Director General is satisfied that any reasonable demand may be met by other means. Although to a certain extent this obligation is superseded in any specific case by entry into a contract with the customer it nevertheless remains in the background and could be transgressed if BT for example consistently failed to provide service or to make it available within a particular time. Moreover the Director General has a duty (under section 3 of the Act) to investigate complaints made with respect to such matters.

It was as a result of a number of complaints that OFTEL had received regarding BT's performance and its standard terms and conditions that it issued its consultative document "PTO Contract Terms and Conditions." This referred specifically to BT's exclusion for any liability for consequential loss arising from any interruption of service or as a result of late delivery. It also referred to the absence of delivery/performance guarantees and invited comment. By the time this book is published, in the light of submissions to him the Director General should at least have issued his report and possibly recommendations as to steps to be taken to meet customer complaints.[14]

One particular solution with which the Director General is apparently toying is a link between BT's performance and its ability to raise prices. There could be difficulties in obtaining sufficiently accurate information to make such a form of regulation fair and effective, unless BT's performance is measured in relation to quality of service statistics either researched and presented by OFTEL itself or prepared by BT but properly audited by OFTEL or on its behalf.

Certainly as networks evolve and become more sophisticated it should be less problematical for PTOs to offer a range of services on different terms and according to different measures of liability for performance. Mercury already does this in relation to its private service contracts and including provision for a rental rebate to customers where service is unavailable for a specified period.

4. Customer Insolvency, etc.

In any case where a company, which is a customer of a PTO, is the subject of an administration order or goes into liquidation or suffers any similar action as listed in section 233 of the Insolvency Act 1986, the PTO cannot make it a condition of continuing service that any outstanding charges are paid, provided the administrator or liquidator has personally guaranteed payment for such service. This requirement also extends to the PTO not doing anything (for example threatening to cut off) to the same effect as such a condition.

E. SUPPLY OF APPARATUS

Apparatus are goods and are therefore assimilated with the general law in relation to sale of goods, particularly the Sale of Goods Act 1979, the Supply of Goods and Services Act 1982 and the Consumer Protection Act 1987.

Conditions implied under the Sale of Goods Act are well known and understood and supported by a considerable body of case law; readers are therefore referred to other works on this specific subject. The Supply of Goods and Services Act ("SGSA") discussed above in relation to liability for the supply of telecommunication services, also deals with the hiring of goods. This is important for suppliers of telecommunication services, particularly the PTOs, as much of their equipment is hired to consumers. The SGSA implies certain conditions as to the right of possession of the goods and, where the owner (bailor) is acting in the course of a business, as to their quality and fitness.

The Consumer Credit Act 1974 ("CCA") regulates "consumer hire agreements" being agreements made by a person with an individual ("the hirer") for the bailment of goods to the hirer. Such an agreement must be capable of subsisting for more than three months and must not require the hirer to make payments exceeding £15,000. If these conjunctive elements are not present the agreement will not be regulated. For these purposes an individual includes a partnership.

Moreover, a telecommunications operator carrying on a consumer hire business of this kind must have a licence under the CCA, obtained by application to the Office of Fair Trading. Public telecommunications operators could apply to the Secretary of State for exemption from the requirements of the Act in relation to consumer hire agreements but so far no such exemption has been conferred.

The form and content of regulated consumer hire agreements is dictated by regulations laid down under the CCA. These requirements are detailed and comprehensive and careful attention has to be given to them; if the regulations are not complied with the agreement will only be enforceable against the hirer by court order.

Under the Consumer Protection Act 1987 ("CPA") strict liability is with certain qualifications, imposed for death or personal injury or loss or damage to "private" property "caused wholly or partly by a defect in the product."[15] The persons liable for such damage can include the producer manufacturer, including any assembler, a person who puts his trade mark or name on a product, and any importer into the EEC. Such liability cannot be excluded by contract.

It is also important to note that under Part II of the CPA a person is guilty of an offence if he supplies goods to or for *consumers* (*i.e.* not to industry or business) which fail to comply with the general safety requirement (*i.e.* which are not "reasonably safe having regard to all the circumstances"). It will also be a criminal offence to supply goods prohibited by safety regulations. All these safety requirements are further explained in the legislation.

F. CONTRACTUAL FORMATION BY TELECOMMUNICATION

The general principle of law applicable to the formation of a contract by offer and acceptance is that the acceptance of the offer by the offeree must be notified to the offeror before a contract can be regarded as concluded: *Carlill* v. *Carbolic Smoke Balls.*[16]

As an exception to this general rule, where a contract is made by letter or telegram the acceptance is deemed to be completed as soon as the letter is put in the post box and that is the place where the contract is made. Identifying the place where the contract is made may be uninteresting for the parties, but in litigation over international contracts it is important: one of the grounds on which the English courts will give leave to serve a writ outside their jurisdiction under Order 11, Rules of the Supreme Court, is if the contract has been made in England.

This so called "postal rule" is based on considerations of practical convenience and commercial expediency, arising from the delay which is inevitable between delivery of a letter or telegram and receipt of it.

However, where a contract is made by instantaneous communication, for example by telephone, the contract is complete only when the acceptance is received by the offeror, and the contract is made at the place where the acceptance is received. The Court of Appeal in *Entores Limited* v. *Miles Far East Corporation*[17] decided that an offer accepted by telex was effectively instantaneous and therefore fell within the general rule and did not therefore fall within the exception covering letters and telegrams.

The *Entores* case was considered by the House of Lords in *Brinkibon*v. *Stahag Stahl.*[18] The House of Lords unanimously approved the *Entores* principle as a general, though not necessarily a universal, rule. Lord Wilberforce said:

> "Since 1955 the use of telex communication has been greatly expanded and there are many variants on it. The senders and recipients may not be the principals to the contemplated contract. They may be servants or agents with limited authority. The message may not reach, or be intended to reach, the designated recipient immediately: messages may be sent out of office hours, or at night, with the intention, or on the assumption, that they will be read at a later time. There may be some error or default at the recipient's end which prevents receipt at the time contemplated and believed in by the sender. The message may have been sent and/or received through machines operated by third person. And many other variants may occur. No universal rule can cover all such cases; they must be resolved by reference to the intentions of the parties, by sound business practice and in some cases by judgment where the risks should lie . . ."

Accordingly where an electronic communication is not instantaneous for any reason, analogies to the postal rule may be apposite.

Regarding communication by facsimile or other forms of communication, the following principles can be drawn from these authorities:

(a) Acceptance of a contract by facsimile will, as a general rule, take effect when the fax is received on the offeror's fax machine. This will also be the place where the contract is made. The fax is, however, analogous with telex in that although this can be stated as a general rule, there may be certain situations, for example when the machines are operated by third parties, where this general rule will not apply. On the whole, however, fax communication should be treated as "instantaneous communication," similar to telex.

(b) Communication using electronic message-handling systems, such as electronic mail is not so staightforward. In some electronic mail systems

> a message may be added to a recipient's mailbox immediately on being sent. In others the message may be delayed and stored in the course of transmission. Clearly when the sender gives the command "send" to the system he has, in ordinary mail language, "posted" the message. However he could delete the message before the addressee can read it or send a further message cancelling the previous message, which could be "collected" by the recipient at the same time as the first message.

At the distant end matters are even more capable of variation. The addressee will not "collect" his message until he "logs in" to the host computer. Even if he does not then "call up" the message he is in the position of having knowingly received mail without, as it were, having opened it.

The essential question is, however, where and when the message is then delivered to the recipient. The prudent approach would be to apply the principles laid down by Lord Wilberforce in *Brinkibon*, namely that no universal rule can cover all cases and that the problem should be resolved by reference to the parties' intentions, sound business practice and a judgment of where the risks should lie.

With electronic mail systems, therefore, in the author's view it is probably safest to assume that delivery would normally take place when the message is entered into the recipient's mail box even though the recipient may then be unaware of its arrival. Unfortunately though this does not deal satisfactorily with the situation where the sender of the message is able subsequently to delete the message or cancel it by a further message. Accordingly, the courts may have to develop special rules for that particular situation and where, for example, the sender is aware that his original message has not been read by the recipient judges may well take the view that in the meantime senders should be able successfully to cancel such messages by subsequent communication.

Since there have been no judicial pronouncements on the latest forms of electronic communication the safe course is for parties actually to specify their proposed modes of communication and, in their contracts, how and when notices are to be taken as received. The growing practice of "registered" electronic mail, whereby this action is notified to the sender, would reinforce this.

Faults could arise within a telecommunication system which prevent the message reaching the addressee's mail box or otherwise corrupt the message received. Where any message, for example by telex or facsimile or electronic mail, is garbled, or in error in some way, the basic legal principle is that the sender of the message is not responsible for the faults of the operator of the communication system or the system itself. In the case *Henkel* v. *Pape*,[19] where the offer was to buy 50 rifles and the defendant, not wanting this number, telegraphed "send 3 rifles," the telegram in fact reached the seller in the form "send the rifles" and the seller therefore despatched 50 rifles. It was held that the buyer was not bound to accept more than 3 rifles.

In the *Entores* case Lord Denning discussed failures in what would now be the recipient's (branch) system. He said:

> "In all instances I have taken so far, the man who sends the message of acceptance knows that it has not been received, or he has reason to know it. So he must repeat it. But suppose that he does not know that his message did not get home. He thinks it has. This may happen if the listener on the telephone does not catch the words of acceptance, but nevertheless does not trouble to ask for them to be repeated: or if the ink on the teleprinter fails at the receiving end, but the clerk does not ask for the message to be repeated: so that the man who sends an acceptance reasonably believes that his message has been received. The offeror in such circumstances is clearly bound, because he will be estopped from saying that he did not receive the message of acceptance. It is his

> own fault that he did not get it. But if there should be a case where the offeror without any fault on his part does not receive the message of acceptance, yet the sender of it reasonably believes it has got home, when it has not, then I think there is no contract."

The point to note here is therefore that if the recipient's system is at fault he is nevertheless to be deemed to have received the message as sent. The same principle would apply to bind the sender where it is his system which is faulty.

The other area which is important to cover in relation to contracts is that of service of notices and other documents, by way of electronic mail. It is important, in particular, where electronic mail is a permitted method of service, to stipulate when the communication is deemed to be made. It is recommended that this should be the date it is first stored in the other party's electronic mailbox, to avoid the possibility of delayed accessing.

Increasingly electronic mail service providers are offering the facility of automatic forwarding of messages from the host computer to customer (*i.e.* addressee) terminals. Where this is the case the legal effects of such a system of conveying messages become very much assimulated with telex, assuming the forwarding is in fact done in "real time".

Notes

[1] S.2 Copyright (Computer Software) Amendment Act 1985; see also clause 161 of the Bill.
[2] See Chap. 2 above, at p. 28.
[3] H.L. Hansard, Vol. 490, No. 43, col. 867 (November 30, 1987).
[4] *Coditel* v. *Cine Vog Films* [1981] 2 C.M.L.R. 362.
[5] COM (84) 300.
[6] s.9.
[7] 75H of C Official Report 156.
[8] See Chap. 7 at p. 122 above.
[10] (1854) 9 Exch. 341.
[11] [1949] 2 K.B. 528.
[12] See s.12 UCTA.
[13] Chap. 4 at p. 48 above.

[14] Director General's statement of March 24, 1988.
[15] s.2 CPA.
[16] [1893] 1 Q.B. 256 at 262.
[17] [1955] 2 All E.R. 493.
[18] [1983] 2 A.C. 34.
[19] (1870) C.R. 6 Exch. 7.

APPENDIX A

(Statement by Minister for Information Technology: November 17, 1983)

The Government have consistently sought to introduce competition into the three main sectors of the telecommunications market—the supply of apparatus for connection to networks, which the hon. Gentleman has been discussing, the provision of services over networks and the running of those networks. The process has been gradual, because we have been determined to avoid market disruption and to give all concerned—BT, its traditional suppliers and new market entrants—an adequate opportunity to adjust as we progress to the competitive environment. Much progress has been made, and the Government have been considering the strengthening of competition in the future.

We have arranged for the making of independent standards under the aegis of the British Standards Institution, and for an independent approvals system under the new British Approvals Board for Telecommunications, which was mentioned by the hon. Gentleman. We have had to start with a clean slate, as no independent standards existed until now. The British Standards Institution is now drafting the independent and objective standards which apparatus must meet before connection to the BT networks, and the work is expected to be substantially completed by the end of 1984. My hon. Friend will shortly announce revised arrangements for BT approvals of products which it sells.

Like the United States of America, the United Kingdom will soon enjoy a completely independent system of approvals for telecommunication apparatus.

I announced in February the Government's intention to introduce competition into the two remaining significant areas of apparatus supply from which it had hitherto been excluded. I can now confirm that from the end of 1984 the BT monopoly of supply and maintenance of the prime telephone instrument connected to the BT systems will end for those customers who have a standard socket fitted by BT. BT have agreed to provide such sockets on demand. Maintenance of newly-installed call-routing apparatus is being progressively liberalised and will be fully open to competition by November 1986.

In short, in the brief period since the British Telecommunications Act came into force in October 1981 we have moved from a position of almost complete monopoly of apparatus supply to a position where a completely open and competitive market is in sight.

I turn now to services provided over public telecommunications networks. The convergence of computing and telecommunications technology has created an entirely new market for these value added network services, or VANS. The VANS general licence authorises anyone who registers with my Department to run telecommunication systems to provide such services over public networks. This is the most liberal VANS regime in the world. The essential condition imposed on those who offer VANS is that they must provide a service over and above the basic telecommunication service of conveying or switching a message. Sixty VANS providers have registered under the general licence their proposals for some 200 different services. Many are very small companies. I expect a continuing rapid development of this sector of the market.

I come to the provision of public networks. It is in that area that we have taken the most far-reaching steps towards competition. The provision of the telecommunications infrastructure has traditionally been the preserve of single monopoly operators throughout the world. In this country, however, we decided on a different course. We encouraged the development of a separate and independent network, Mercury Communications Limited, which has been licensed to provide every form of digital telecommunication service, including private leased circuits, switching services to business and domestic premises, and the full range of international services. We expect Mercury to develop into a national public telecommunication system and we intend the new licence to be granted to Mercury under the Telecommunications Bill to contain specific obligations about the installation of the Mercury network on a national basis having regard to Mercury's commercial plans. Mercury will have a licence similar to BT's.

The creation of a new network requires very large investment, which will mean a long period before the investment can yield a return. Mercury therefore needs time to install and consolidate its national network. Similarly, British Telecom also needs time to adjust. Its public telecommunications system was developed as a unitary and integrated network without thought of competition and it needs time to adapt to competition.

Telecommunications can be provided only by means of cables or radio. The installation of cables may have undesirable effects on the environment and disrupt highway surfaces and traffic. The radio spectrum is a limited resource, with many competing demands on it. It would not be right to license a multiplicity of operators in any one area to install overhead wires or to dig up streets, nor would it be practicable from a radio spectrum standpoint, at least for some time to come, to license more than two national public telecommunication networks. The Government therefore gave assurances when the Mercury licence was first granted in 1982 that for the foreseeable future we would not license any additional national public telecommunications network.

To avoid uncertainty the Government have now decided to make it clear that we do not intend to license operators other than BT and Mercury Communications Limited to provide the basic telecommunication service of conveying messages over fixed links, whether cable, radio or satellite, both domestically and internationally during the seven years following this statement. The position will then be reviewed. In the Hull area, the city of Kingston upon Hull will continue to exercise the functions of public telecommunication operator.

For telecommunication systems serving specialised

market segments, Government policy is as liberal as practicable. The Government have already licensed two national cellular mobile radio telephone networks. They will provide increasing alternatives to BT and Mercury at the local level but their role will be limited to providing mobile radio telecommunication services.

The Government will shortly be licensing the first new broadband cable networks and, as the cable White Paper made clear, cable companies will be licensed to offer a full range of telecommunication services within their licensed areas, but only BT or Mercury will be permitted to offer voice telephony services on cable systems either alone or in partnership with the cable companies.

The Government will also be exploring the scope for introducing further mobile radio services when the bands I and III frequencies become available in 1985. Moreover, the Government will keep under consideration ways of introducing new specialised services by satellite. The cellular radio companies, the cable operators and other operators, including private mobile radio, will be required to obtain the fixed links connecting their systems to other systems only from BT, Hull and Mercury.

Arrangements for interconnecting different telecommunication systems are particularly important. The Government intend that any subscriber to one public telecommunication system should be able to call any subscriber to other public telecommunication systems. The Telecommunications Bill and the licences to be granted under it will provide Mercury with a right to compete equally, and to interconnect with, BT. The draft BT licence published on 25 October contains an obligation requiring BT to connect, or to permit the connection of its systems to any other system where the operator of the other system is licensed to connect and requires the connection. Hon. Members are familiar with the complexity of the issue, but that is the obligation imposed on BT in the licence.

A similar obligation will be contained in the new Mercury and Hull licences. It will be for BT, Mercury and Hull to negotiate the technical and commercial terms for the connections both between their systems and between their systems and other systems. Connection may be on standard published terms where applicable or by agreement between the operators. Where there is no agreement, those concerned will be obliged to carry out connections on terms and conditions laid down by the Director General of Telecommunications. The methods and arrangements by which connection can be effected are complex and will be the subject of future discussions with the relevant parties.

The BT licence leaves open the possibility of the future introduction of various forms of "resale" of leased circuits; so will the new Mercury and Hull licences. The Government have decided to explore the scope for reducing some of the restrictions currently imposed on the use of circuits leased from BT and Mercury. We are, therefore, discussing the possibility of allowing group use of inland leased circuits, and some easing of current restrictions on the interconnection of leased circuits and the public switched networks. The Government expect to reach conclusions on these issues in the new year.

As regards "resale" of international circuits, the Government stand by current international agreements. We are also considering whether appropriate terms and safeguards can be devised to allow BT and Mercury to bring into public use in their networks spare capacity which may be available on privately-owned networks.

Some forms of "resale" and satellite services, combined with interconnection with the public switched networks, could develop the characteristics of an additional network. The Government wish to emphasise that any developments in policy on "resale" or satellite services will be consistent with the policy I have outlined today, which is not to license anyone other than BT and Mercury to run national public telecommunication networks until November 1990 at the earliest. Other forms of resale than those I have just described will not be licensed in the period before July 1989, which corresponds to the period in the BT licence for the RPI minus X formula.

It will be obvious to those who have followed what I have been saying that this is a complicated matter. I have pulled together the threads of the competition policy which has been developing over the past two to three years. Our liberalisation policy aims not only to increase consumer choice but to stimulate greater efficiency in the use of natural resources and encourage growth and innovation. The competitive framework that I have outlined will enable those aims to be achieved and will allow BT, Mercury and all the other companies in this rapidly expanding industry every opportunity to prosper through fair competition to the benefit of consumers, their work forces and the economy generally.

APPENDIX B

PUBLIC TELECOMMUNICATIONS OPERATORS

1. Public Switched Networks

National
British Telecommunications Plc
Mercury Communications Limited

Hull
Hull City Council/Kingston Communications (Hull) Plc

2. Cable TV

Aberdeen Cable Services Limited and Broadband Ventures Limited
British Cable Services Limited
CableTel Communications Limited
Clyde Cablevision Limited
Coventry Cable Limited and Broadband Ventures Limited
Croydon Cable Television Limited
East London Telecommunications Limited
Swindon Cable Limited
Westminster Cable Company Limited and Broadband Ventures Limited
Windsor Television Limited

3. Cellular Radio

Racal Vodafone Limited
Telecom Securicor Cellular Radio Limited

INDEX